TERTIARY LEVEL BIOLOGY

The Lichen-Forming Fungi

DAVID L. HAWKSWORTH, Ph.D., D.Sc., F.L.S., F.I. Biol.
Director
Commonwealth Mycological Institute
Kew

DAVID J. HILL, D.Phil.
Tutor in Biology
Department of Extra-Mural Studies
University of Bristol

Blackie
Glasgow and London

Distributed in the USA by
Chapman and Hall
New York

Blackie & Son Limited
Bishopbriggs
Glasgow G64 2NZ

Furnival House
14–18 High Holborn
London WC1V 6BX

Distributed in the USA by
Chapman and Hall
in association with Methuen, Inc.
733 Third Avenue, New York, N.Y. 10017

British Library Cataloguing in Publication Data
Hawksworth, D.L.
  The lichen-forming fungi.—(Tertiary
  level biology)
  1. Lichens
  I. Title   II. Hill, David J.   III. Series
  589.1   QK583
  ISBN 0–216–91633–X
  ISBN 0–216–91634–8 Pbk

Library of Congress Cataloging in Publication Data
Hawksworth, D.L.
  The lichen-forming fungi.

  (Tertiary level biology series)
  Bibliography: p. 142
  Includes index.
  1. Lichens.   2. Fungi.   I. Hill, David J.
  II. Title.   III. Series.
  QK583.H39   1984   589.1   84–1020
  ISBN 0–412–00631–6
  ISBN 0–412–00641–3 (pbk.)

Printed in Great Britain by
McCorquodale (Scotland) Ltd., Glasgow

Preface

Lichen associations include some of the oldest living organisms and represent a major nutritional method adopted by one in five fungi. Major advances in our knowledge of these biologically fascinating organisms have been made in recent years and they now have a great deal to offer to teaching in colleges and universities. In addition to being examples of biotrophic systems, they merit discussion in courses on fungal phylogeny, fungal nutrition, ecology, ecophysiology, biogeography, evolution, chemotaxonomy, environmental monitoring, and algology. As all aspects of lichenology cannot be treated adequately in a book of this length, we have emphasized topics which we have found to be of particular interest to a advanced undergraduate and postgraduate biologists (or biogeographers) or those contemplating more detailed studies in particular areas. Consequently we have endeavoured to place lichen associations in the broader context of biological and biogeographical teaching.

Examples are drawn from many areas of the world, including North America, but it is inevitable that European ones predominate, as lichens there are better known than in other regions. Technical terms are defined when first used, and definitions can therefore be located with the aid of the index.

Lichens are not a systematic group and so are not appropriately treated in a groups-orientated programme, but are a major biological phenomenon all too commonly accorded scant attention in university courses. While this book can also be employed as the basis of special courses devoted to lichens, it is our primary intention that it will be utilized much more widely and contribute to the integration of lichenology in mainstream biology teaching. This is long overdue, but we are confident that teachers and lecturers will find it a mutually beneficial symbiosis.

In preparing this text we have benefited from the comments of several colleagues, including Dr D.H. Brown, Mr P.W. James, Professor D.H. Lewis, Dr M. Madelin, Dr F. Rose and Dr M.R.D. Seaward. For making originals of illustrations available we are also indebted to Professor V. Ahmadjian, Mr F.S. Dobson, Dr M. Galun, Dr R. Honegger, Dr D. Jones, Dr A. Pentecost, and Miss F.J. Walker.

D.L.H.
D.J.H.

TERTIARY LEVEL BIOLOGY

A series covering selected areas of biology at advanced undergraduate level. While designed specifically for course options at this level within Universities and Polytechnics, the series will be of great value to specialists and research workers in other fields who require a knowledge of the essentials of a subject.

Recent titles in the series:

Locomotion of Animals	Alexander
Animal Energetics	Brafield and Llewellyn
Biology of Reptiles	Spellerberg
Biology of Fishes	Bone and Marshall
Mammal Ecology	Delany
Virology of Flowering Plants	Stevens
Evolutionary Principles	Calow
Saltmarsh Ecology	Long and Mason
Tropical Rain Forest Ecology	Mabberley
Avian Ecology	Perrins and Birkhead

In the USA, these titles are distributed by
Chapman and Hall, New York

Contents

Acknowledgments

The following illustrations are reproduced by permission:
Figure 2.8, p. 25, from *Biblioteca lich*. 1: 1–123 (1973).
Figure 3.3, p. 33, from Henssen/Jahns, *Lichenes*, Georg Thieme Verlag, Stuttgart, 1974.
Figure 3.5, p. 35, from *J. Hattori Bot. Lab*. 52: 419 (1982).
Figure 4.2, p. 47, reprinted by permission from *Nature* 289: 169–172, © 1981, Macmillan Journals Ltd.
Figures 7.4–7.6, pp. 108–9, 111–13, from Seaward and Hitch, *Atlas of the Lichens of the British Isles*, by permission of Biological Records Centre, Institute of Terrestrial Biology (Natural Environment Research Council), 1982.

CHAPTER ONE

THE LICHEN HABIT

1.1   History

'Lichen' (*lie 'ken*) was introduced into Greek literature in about 300 BC by Theophrastus, primarily to describe outgrowths from the bark of olive trees. Some lichens were already being exploited and were included in the earliest *herbals*, accounts of medically useful plants appearing from the early sixteenth century. [The same word was also used for outgrowths from human tissues and about 10 'lichen' disease names (e.g. 'lichen planus') are currently in use in medicine.] The French botanist J.P. de Tournefort (1656–1708) used 'lichen' as a generic name (*Les Élémens de Botanique*, 1694), but it was the Italian P.A. Micheli (1679–1737) who first introduced an ordered system (*Nova Plantarum Genera*, 1729). Micheli enumerated about 300 species within 38 'orders', illustrated asci for the first time in any fungus (figure 1.1), and observed the development of lichen thalli from soredia.

A Swedish doctor, Erik Acharius (1757–1819), must be credited as the founder of the systematic study of lichens. He introduced many of the descriptive terms and generic names now used, and synthesized the extant world literature (*Methodus Lichenum*, 1803; *Lichenographia Universalis*, 1810; and *Synopsis Methodica Lichenum*, 1814). Acharius largely ignored

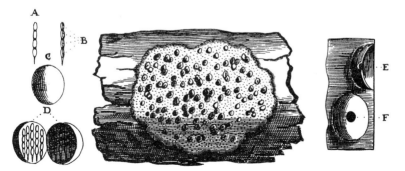

Figure 1.1   Drawings of a *Pertusaria* species from Micheli (1729), *Nova Plantarum Genera*, showing asci and included ascospores.

1

microscopic characters. Their importance was grasped by G. de Notaris (1805–77), G.W. Körber (1817–85), A.B. Massalongo (1824–60), and others, who described a succession of new genera and species. The studies of these lichenologists in the mid-nineteenth century, the significance of which is only now being appreciated, became eclipsed by the work of W. Nylander (1822–99). Nylander published prolifically, making many valuable contributions to the subject. However, he never accepted the dual nature of lichens, which had been elucidated by S. Schwendener (1829–1919) in 1867, and adopted a rigid classification system ignoring crucial microscopic characters. The studies of Körber and his contemporaries were further obscured as the compendia of lichen families, genera and species prepared by Alexander Zahlbruckner (1860–1938) in Vienna were based on Nylander's concepts. Zahlbruckner's listings remain vital reference works (e.g. *Catalogus lichenum universalis*, 10 vols., 1921–40).

Studies on chemicals produced in lichens advanced rapidly in the 1930s (see Chapter 8), but in contrast physiological studies were rather late in developing. Not until 1961 was carbon transfer between the partners conclusively demonstrated by David C. Smith. Critical ecophysiological studies date from the independent investigations of Otto L. Lange and A. Ried in 1953. Concern for the environment has led to the use of lichens in the monitoring of pollutants; the first zonal scale calibrated according to sulphur dioxide levels was published in 1968 by Oliver L. Gilbert.

Ever since the pioneering studies of J.A. Nannfeldt in 1932 and E.S. Luttrell in 1951 on the classification of ascomycete fungi as a whole, a gradual move towards the development of a comprehensive system of ascomycete orders and families encompassing both lichenized and non-lichenized fungi has been taking place. Lichen-forming fungi have been included in both the *Index of Fungi* and *Dictionary of the Fungi* from 1971, and were eliminated as a 'group' from the International Code of Botanical Nomenclature in 1981.

Integration is a key word in lichenology today. Carbohydrate transfer in lichens has been found to be similar to that of other mutualistic symbioses (including mycorrhizas and alga–invertebrate symbioses), and increased attention is being accorded to the role of lichens in ecosystems.

1.2 Defining 'lichen'

'A lichen is an association of a fungus and a photosynthetic symbiont resulting in a stable thallus of specific structure.' This definition was the winner in a poll the International Association for Lichenology held amongst its members in 1981. However, what constitutes a 'thallus of specific structure' remained a matter for personal opinion. This leads to the exclusion of some structurally very simple associations which may be physiologically comparable to more structurally complex ones. An alternative definition, 'a stable self-supporting association of a mycobiont and a photobiont', was adopted in the 1983 edition of the *Dictionary of the Fungi* and recommended for general use. The *mycobiont* (the exhabitant) is the fungal partner in a lichen association, and the *photobiont* (the inhabitant) the photosynthetic partner. Interestingly, if the word 'phycobiont' were replaced by 'autotrophic vascular plant' the above definition could equally be applied to mycorrhizas. Photobionts may be either green algae (phycobionts) or cyanobacteria (cyanobionts); cyanobacteria were formerly often referred to as blue-green algae.

A definition in biochemical or physiological terms is not possible as too few critical associations have been examined by such techniques, but lichen associations often arise when the bionts are united in what to most organisms would be adverse environmental situations.

The difficulty in arriving at a generally acceptable definition of 'lichen' arises because of the variety of fungal–algal associations and the range of biological situations to be seen amongst *symbioses* (unlike organisms living together) studied by lichenologists.

1.3 Fungus–alga associations

Fungi and algae can enter into a wide spectrum of biological relationships. An appreciation of this situation is fundamental to discussions of the lichen mode of life and the nature of the partners forming lichens. These relationships can be divided into two major nutritional types; (1) *antagonistic* (or necrotrophic) interactions where the fungi involved take nutrients from the photobionts and eventually kill them, and (2) *mutualistic* (or biotrophic) interactions where the fungus forms a stable association with the photobiont and neither partner is eliminated. In mutualistic symbioses the reproductive fitness of both partners increases as a result of the interaction.

The antagonistic mode of life is the commonest in the Fungi as a whole. Saprophytic fungi living on dead animal and plant materials, and parasitic or pathogenic fungi causing diseases and often death in their hosts all belong in this category. Antagonistic fungus–alga interactions do occur; for example, about 120 different fungi are restricted to marine algae (especially larger seaweeds). The biological status of many is far from clear, but this group encompasses major pathogens such as *Chadefaudia gymnogongri*, which has a considerable host range, and also saprophytes, for example *Corallospora maritima* on dead fucoids and red algae. In the freshwater environment and on land, algae are also attacked by fungi. Numerous members of the Chytridiales have this lifestyle, for instance *Chytridium olla* parasitizing *Oedogonium* species, and the hymenomycete *Athelia epiphylla* causing massive brownish lesions in colonies of *Desmococcus* and also other pleurococcoid green algal films and lichens in urban areas (figure 1.3*A*).

Some fungus–alga associations which are not clearly antagonistic and which could well be mutualistic are lichen-like in some respects, but are not normally considered by lichenologists. In *Mycosphaerella ascophyllii* (figure 1.2*A*), the fungus mycelium spreads throughout the tissues of the host seaweeds (*Ascophyllum nodosum* and *Pelvetia canaliculata*) and perithecia are scattered over their surface: these seaweeds, at least in Europe, invariably support this fungus. The infected algae are healthy and, as the fungus is systemic, a debate as to whether such associations should be regarded as lichens has continued since the early decades of this century. A special term, *mycophycobiosis*, was coined in 1972 for permanent symbiotic associations between marine algae and fungi in which the alga pre-dominated and retained its normal outward appearance. However, some other associations cannot easily be accommodated within the concept of mycophycobiosis. The fungus *Chadefaudia corallinarium* forms loose

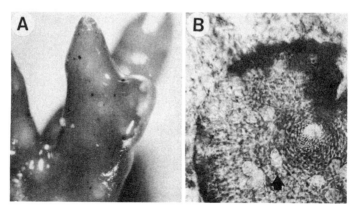

Figure 1.2 *A, Mycosphaerella ascophylii* on *Pelvetia canaliculata* (note black perithecia), × 10. *B, Nectria phycophora*, surface view of perithecium on *Dawsonia grandis* (note locules containing algae, arrow), × 100.

associations with various marine algae which have been interpreted as primitive lichens, and *Arthopyrenia pelvetiae* on *Pelvetia canaliculata* even has blue-green photobiont cells intimately associated with it. Comparable situations are now also documented in the terrestrial environment. *Phycorella scytonematis* infects single cells in filaments of the blue-green *Scytonema* by haustoria in what has been termed 'biotrophic parasitism'. *Orbilia luteorubella* forms an association with an *Anacystis*-like photobiont, and *Cudoniella brasiliensis* and *Pezizella parasitica* with unicellular green algae; the extent of specialization in these and comparable examples is certainly no less than some taxa traditionally regarded as lichens, for instance *Dimerella diluta* (figure 2.1*D*) and *Micarea prasina*.

Studies on the bright green algal growth so common on shaded nutrient-rich bark in urban areas show that it is also often a lichen-like association. A dematiaceous hyphomycete, *Coniosporium aeroalgicola*, appears to be involved, judging from studies carried out recently in Switzerland. This same association can act as a pioneer on granitic rocks.

There are other cases where fungi appear to be able to form associations with algae on an even more casual basis, for example in the brown felt blight fungus *Herpotrichia juniperi*. The establishment of such facultative associations parallels those seen in some fungi traditionally studied by lichenologists but in which algae are difficult or impossible to find; this situation is especially characteristic of perithecioid lichens on smooth bark such as some species of *Arthopyrenia*, *Leptorhaphis* and *Mycoporum*. *Stenocybe septata*, long regarded as a lichen, is now known not to be intimately associated with any particular algae, but often grows through surface algal colonies from inside the underlying bark.

Fungus–alga associations can also sometimes involve other kinds of plants. Of particular note are cases involving bryophytes. *Nectria phycophora*, on leaves of *Dawsonia grandis*, has packets of algal cells in special locules in the perithecial wall (figure 1.2*B*). *Arthopyrenia endobrya* is associated with filamentous green algae living *within* the leaf cells of Brazilian hepatics of the Lejeuneaceae, and the thallus of *Vezdaea aestivalis* not only develops subcuticularly with its algae on various bryophytes, but the hyphae of the mycobiont can kill the host leaves.

1.4 Three- and four-membered symbioses

In addition to the above, more complex relationships exist. Two fungi may associate with a single photobiont. This situation occurs in the *lichenicolous* (lichen-inhabiting) *fungi*; there may be 300 genera and 1000 species of obligately lichenicolous fungi and between them there is a wide range of biological relationships (figure 1.3). About half the genera involved only include lichenicolous species, the remainder may also encompass other nutritional types (see table 1.3, section 1.5). A few occur as saprophytes on dead or decaying lichen thalli (e.g. *Niesslia cladoniicola*), but most are parasitic, causing necrotic patches (e.g. *Lichenoconium lecanorae*, figure 1.3*B*; *Phoma cytospora*), killing lobes or substantial parts of thalli (e.g. *Lichenoconium erodens, Nectriella tincta*, figure 1.3*C*), or giving rise to gall-like deformations (e.g. *Polycoccum galligenum*, figure 1.3*E*); this last situation is also of particular morphogenetic interest (see section 2.6). A stable situation in which the host lichen is, from outward appearances, totally unaffected by the presence of a lichenicolous fungus is termed

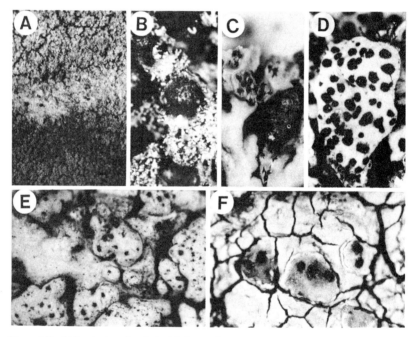

Figure 1.3 Lichenicolous fungi. *A, Athelia arachnoidea* advancing over and killing *Lecanora conizaeoides* (dead lichen at the top), × 10. *B, Lichenoconium lecanorae* on *Parmelia saxatilis*, a single pycnidium is included in each infection spot, × 10. *C, Nectriella tincta* on *Anaptychia fusca* causing bleaching, × 10. *D, Nesolechia oxyspora* on *P. saxatilis* causing bleaching and deformations, × 10. *E, Polycoccum galligenum* on *Physcia caesia* causing gall-like outgrowths including the perithecia, × 20. *F, Arthonia glaucomaria* living parasymbiotically in the apothecia of *Lecanora rupicola*, × 20.

parasymbiosis (i.e. a symbiotic association with an already existing symbiosis); parasymbionts are presumed to obtain their carbohydrate requirements from the photobiont in the lichen thallus—physiologically they may therefore be indistinguishable from their hosts. Examples include *Arthonia glaucomaria* on *Lecanora rupicola* (figure 1.3F), and both *Dactylospora parasitica* and *Sphinctrina turbinata* on *Pertusaria* species.

The biological nature of relationships can also change with age. In what may be a not uncommon method by which lichen thalli can be produced, an invading fungus is initially parasitic on a host lichen thallus, killing the mycobiont of the host, but proceeding to usurp its algae and utilize them in the production of an independent two-membered thallus. This situation occurs in *Arthrorhaphis citrinella* on *Baeomyces rufus, Blarneya hibernica* (figure 1.5) on *Enterographa* and *Lecanactis, Chaenothecopsis consociata* on *Chaenotheca chrysocephala,* and *Diploschistes muscorum* on *Cladonia.* This type of parasitism, which results in a different association occupying precisely the site of another, recalls that seen in strangling figs such as *Ficus benjamina* which kill and take the place of the tree on which they first grow.

In contrast to the above examples, a single mycobiont can form an association with two (exceptionally three) different photobionts. This is seen in the development of structures termed *cephalodia* which are often an integral part of the lichen or a special outgrowth from it; these are discussed further in section 2.6 where their morphogenetic importance is also considered. The specificity of photobionts is reviewed separately in section 1.6.

In 1977 Poelt proposed that associations involving three partners should be termed *three-membered symbioses* (1 mycobiont and 2 photobionts; *or* 2 mycobionts and 1 photobiont) and ones with four partners *four-membered symbioses* (2 mycobionts with 2 photobionts; *or* 1 mycobiont with 3 photobionts). The range of situations known is illustrated in figure 1.4.

The *lichenicolous lichens* (lichen-inhabiting lichens) are perhaps the best examples of four-membered symbioses. In these each mycobiont has an independent photobiont. In some cases the nature of such associations is readily recognizable as the thalli differ in colour or position, as in *Lecidea insularis* on *Lecanora rupicola* and *Verrucaria aspicilicola* on *Aspicilia calcarea.* However, in others it can be difficult to appreciate as with *Buellia pulverulenta* which has its photobiont cells located *within* the foliose thallus of its host (*Physconia pulverulacea*) with only the ascomata of the *Buellia* visible from above; not unsurprisingly, *B. pulverulenta* was not recognized as lichenized until 1980—it had been interpreted as a lichenicolous fungus for the previous 120 years.

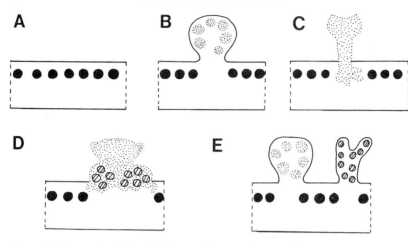

Figure 1.4 Types of symbioses in lichens. *A*, Two-membered (1 photobiont + 1 mycobiont; lichen). *B*, Three-membered (2 photobionts + 1 mycobiont; lichen with cephalodia). *C*, Three-membered (1 photobiont + 2 mycobionts; lichen with parasymbiotic fungus). *D*, Four-membered (2 photobionts + 2 mycobionts; lichen with lichenicolous lichen). *E*, Four-membered (3 photobionts + 1 mycobiont; lichen with cephalodia containing two different photobionts).

1.5   Mycobionts in lichens

The formation of a lichen association can be seen as one of the available strategies by which a fungus can satisfy its need for carbohydrates required for respiration and growth. The effectiveness of the method is evidenced by the fact that one in five of the 64 200 known fungi is lichenized. Alternative strategies are the development of parasitism, saprophytism, or other stable mutualistic associations such as mycorrhizas. It is important to recognize lichenization in this way, as a life-style equivalent to parasitism or saprophytism and not as something inherently different.

If lichenization was a nutritional option, it would not be unreasonable to expect different groups of fungi to have independently developed the ability to form lichen associations; this proves to be so. Indeed, the broad spectrum of mutualistic fungus–alga associations and the difficulty of coining a clean-cut definition of 'lichen' itself largely arises because lichenization has reached a higher level of complexity in some groups than in others.

The number of lichenized fungi is about 13 500*, but 98% of these (13 250

---

*Figures in the range 15 000–20 000 are commonly cited but are now known to be excessive.

Table 1.1 Orders of Ascomycotina including lichen-forming fungi.

Arthoniales*	Opegraphales*
Caliciales*	Ostropales*
Dothideales*	Peltigerales
Graphidales	Pertusariales
Gyalectales	Pyrenulales*
Helotiales*	Sphaeriales*
Lecanidiales*	Teloschistales
Lecanorales*	Verrucariales*

* Orders also including non-lichenized representatives

species) belong to the Ascomycotina (ascomycetes). That is the largest major grouping within the fungi, comprising about 2720 genera and 28 650 species dispersed through 37 orders. Sixteen of the orders include, or consist entirely of, lichenized fungi (table 1.1); 46 per cent of all species in the Ascomycotina are lichen-forming.

Only five orders within the Ascomycotina consist entirely of lichenized species: the Graphidales, Gyalectales, Peltigerales, Pertusariales and Teloschistales. All probably originated early in the evolution of the class. In the remaining orders which include lichen-forming taxa, some are predominantly lichen-forming (e.g. Arthoniales, Caliciales, Lecanorales, Opegraphales, Verrucariales), while in others the lichenized taxa are proportionately few (e.g. Dothideales, Helotiales, Ostropales). Considerations of ascus structure, ontogeny, and biogeography strongly suggest that orders in the former category are primarily lichenized and that the saprophytic or parasitic taxa in them are descended from lichen-forming ones. Conversely, in orders such as the Helotiales, the lichen habit appears to be evolving, arising in species scattered through a variety of genera. About twenty genera are now know to include species, some of which are lichen-forming and others of which are saprophytic, parasitic or lichenicolous (table 1.2); these genera can be viewed as crossing biological boundaries. In some cases a genus encompasses species with more than two nutritional modes (see table 1.3).

Lichen associations are much rarer in other major groups of the fungi. This is perhaps not surprising as there is evidence to suggest that the Ascomycotina are much older than the Basidiomycotina, and the evolution of a nutritional method novel for a group cannot be expected to be rapid. Nevertheless, several orders of Basidiomycotina, all in the Hymeno-mycetes, include lichen-forming taxa.

Table 1.2 Examples of genera including lichen-forming and non-lichenized species.

Arthonia	Chaenothecopsis	Opegrapha
Arthopyrenia	Cyphelium	Orbilia
Arthothelium	Diplotomma	Pezizella
Arthrorhaphis	Lecidea	Rhizocarpon
Bacidia	Melaspilea	Sphinctrina
Calicium	Mycomicrothelia	Thelocarpon
Catillaria	Omphalina	Xylographa

In some *Omphalina* (Agaricales, Tricholomataceae) species the mushroom-like fruit bodies arise from lichenized squamules ('*Coriscium*') or granular structures ('*Botrydina*'), but in others different nutritional types occur (table 1.3). Rather loose associations occur in some *Multiclavula* (Aphyllophorales, Clavariaceae) species whose fruit bodies arise from a slimy greenish photobiont crust; interestingly, other species in this genus can form associations with slime moulds and mosses. Perhaps the best-known genus of lichen-forming Basidiomycotina is the exclusively lichenized *Dictyonema* (Aphyllophorales, Thelephoraceae); most of its five species are tropical, several forming substantial brackets (e.g. *D. pavonia*). In view of their rarity, basidiomycete lichens are not considered in detail in the following chapters; general mycological texts can be consulted for details of structure and terminology used in treating these fungi.

Lichenized representatives of the Deuteromycotina ('Fungi imperfecti'), fungi which reproduce by asexual conidia, have only recently started to be studied critically and many (especially from the tropics) await description. So far, 41 genera and 55 species of lichen-forming deuteromycetes have been described, including representatives of both the Hyphomycetes in which conidia arise from hyphae (e.g. *Blarneya hibernica*; figure 1.5) and Coelomycetes in which conidia are formed in special structures called *conidiomata* (e.g. *Lyromma nectandrae*). Numerous lichens forming sexual ascomata (see section 3.1) also have conidiomata (pycnidia; see section 3.6), which in some cases commonly occur without the ascomatal stage.

In 10 lichen genera no fruiting bodies are known; these have often been classified with the Deuteromycotina, but while the position of some is obscure (e.g. *Racodium, Cystocoleus*, figure 2.5), many can now be referred to other orders on the basis of chemical or other characteristics (e.g. *Lepraria, Leprocaulon, Leproplaca, Thamnolia*).

Mention should also be made of the enigmatic *Geosiphon pyriforme* which has *Nostoc* photobiont cells enclosed in specialized globose vesicles.

Table 1.3 Examples of genera including species crossing biological boundaries, i.e. exhibiting different nutritional types.

	Arthonia	Buellia	Chaenothecopsis	Dactylospora	Lecidea	Nectriella	Omphalina	Pezizella
Lichen-forming	tumidula	disciformis	lignicola	—	fuscoatra	—	luteovitellina	parasitica
Lichenicolous lichen	—	pulverulenta	—	—	insularis	—	—	—
Parasymbiont	intexta	—	epithallina	parasitica	supersparsa	robergei	—	epithallina
Parasite	fuscopurpurea	—	parasitaster	saxatilis	vitellinaria	tincta	cupulatoides	—
Fungicolous	—	—	caespitosa	—	—	—	—	—
Algicolous	—	—	—	—	—	laminariae	—	—
Bryophilous	—	—	—	scapanaria	—	—	sphagnicola	polytrichi
Plant mycorrhizae	—	—	—	—	—	—	—	ericae
Saprobe	—	—	debilis	stygia	—	fuckelii	chrysophylla	eburnea

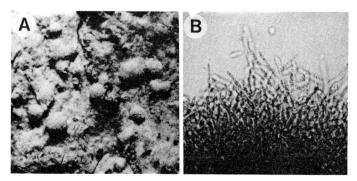

Figure 1.5 *Blarneya hibernica. A,* Sporodochia arising from the lichenized thallus, × 2. *B,* Conidiophores and conidia that comprise the sporodochia, × 325.

This taxon, which occurs with bryophytes on soil in central Europe, most probably belongs to the Mastigomycotina.

Finally, it is interesting to note that there are a few cases of myxomycete-algal ('*myxolichen*') and actinomycete-algal ('*actinolichen*') associations which have been reported but these have not so far been investigated critically.

1.6   Photobionts in lichens

The photosynthetic partners in lichens, with the exception of a few special associations (see section 1.3), are either blue-green Cyanobacteria (Cyano-phyta) or green Chlorophyta. Many are difficult to name reliably even as far as genus without isolation and study in pure culture; normally filamentous types can become almost unicellular within a thallus and the sexual reproductive stages may be suppressed. The practical difficulties, together with the rather uncertain taxonomy of most groups of both Cyanobacteria and terrestrial Chlorophyta, have resulted in a most inadequate knowledge of this aspect of lichen associations.

Some 37 genera have now been identified as lichen photobionts (table 1.4)—a figure in stark contrast to the numbers of mycobionts involved. Further, many of the genera involved are well known outside lichens (e.g. *Nostoc, Trentepohlia*), and few are known only from lichen associations (e.g. *Elliptochloris*). Nevertheless, some of the commonest photobiont genera are hardly known outside lichen thalli (e.g. *Myrmecia, Pseudotrebouxia, Trebouxia*) although there are now several reports of them in the free-living state. Some such colonies may, however, have been

Table 1.4 Genera of photobionts reported as entering into lichen associations.

Chlorophyta		Cyanobacteria
Cephaleuros	Hyalococcus	Anabaena
Chlorosarcina	Leptosira	Anacystis
Chlorella	Myrmecia	Aphanocapsa
Chlorèllopsis	Phycopeltis	Calothrix
? Chlorococcum	? Protococcus	Chroococcus
Coccobotrys	Pseudochlorella	Dichothrix
Coccomyxa	Pseudopleurococcus	Fischerella
Dictyochloropsis	Pseudotrebouxia	Gloeocapsa
Dilabifilium	Stichococcus	Hyella
Elliptochloris	Trebouxia	Hyphomorpha
Gloeocystis	Trentepohlia	Nostoc
Heterococcus	Trochiscia	Scytonema
		Stigonema

derived from soredia or other lichenized propagules and may be ephemeral.

The same algal species can occur in several taxonomically distant lichen genera. For example, *Trentepohlia umbrina* is known from some *Arthonia, Chaenotheca, Graphis* and *Opegrapha* species, and *Trebouxia glomerata* from members of *Cladonia, Huilia* and *Stereocaulon. Chaenotheca carthusiae* can utilize two genera of green algae, *Trebouxia* in Costa Rica and *Stichococcus* elsewhere in the Northern Hemisphere. There are also cases where a mycobiont can form an association with different species from the same photobiont genus, as in *Xanthoria parietina* which can contain either *Trebouxia albulescens* or *T. decolorans*, apparently with no effect on thallus morphology or chemistry.

Pioneering experiments by Ahmadjian and his co-workers in the last few years have tested the ability of different mycobionts to associate with different algal genera and species in culture. For example, a single *Cladonia cristatella* mycobiont isolate was found to be able to re-synthesize with 13 *Trebouxia* isolates but not 10 isolates of *Pseudotrebouxia*. In this set of experiments there was no effect on the secondary metabolites produced nor, in the early stages attained, on thallus form (even at the ultrastructural level).

In some instances, the interaction of a single mycobiont with a green instead of a blue-green photobiont can lead to quite a different morphological structure. This situation has considerable implications for lichen morphogenesis (see section 2.6).

Within the thallus, green photobionts generally reproduce asexually, as by the production of mitotic *aplanospores*, but zoosporogenesis has been documented for *Trebouxia* within the lichen thallus; this latter phenomenon may often be overlooked.

1.7 Taxonomic concepts

Under the rules of the International Code of Botanical Nomenclature, the names given to lichens arȷ treated as referring to the mycobiont; the photobionts can consequently retain independent generic and specific names. It is not therefore surprising that it is the characters derived from the mycobiont that have been given most weight in the recognition of orders, families and genera of lichens. The characters used to separate those categories are now almost exclusively mycological, i.e. ontogeny of the ascomata, ascomatal structure, ascus structure, ascospore type, conidiomatal features. Thallus characters have continued to be used in recent years, especially when anatomical details have become clearer with the use of the scanning electron microscope (SEM), but genera largely based on such features have not found general acceptance; instances are the segregation of *Cladina* from *Cladonia*, of *Fistulariella* and *Niebla* from *Ramalina*, and of *Bulbothrix*, *Hypotrachyna*, *Parmelina*, *Parmotrema*, *Pseudoparmelia*, *Relicina* and *Xanthoparmelia* from *Parmelia*.

Until 30 years ago, genus concepts had in many cases been based on single criteria such as ascospore colour and septation which cannot be condoned in mycological thinking. Indeed, that approach even led to taxa which are now considered as belonging to different orders being included in a single genus, as in the case of *Arthopyrenia* s.lat. and *Microthelia* s.lat. The result of increasing attention to ascomatal ontogeny and ascus structure has been a substantial remodelling of generic limits which, while now well-advanced in some groups, is still far from complete.

While mycological features predominate at the higher levels, thallus features join ones such as dimensions of ascomatal tissues, spores, or spore septation to provide species concepts. Chemical characters also commonly support differentiations based on small morphological discontinuities. Ascomatal and chemical features are in general much more stable as taxonomic criteria than thallus form at species level as the latter can be affected by environmental constraints (see section 2.2). Before deciding that a discontinuity that might merit species separation exists it is consequently advisable to study large numbers of individuals to ensure that the samples are not a part of a continuum. For example, in monographing the North American species of *Alectoria*, *Bryoria* and allied genera, Brodo and Hawksworth studied almost 9000 specimens. Inadequate sampling has led to the description of numerous taxonomically worthless taxa, especially in crustose genera such as *Acarospora* and in the Lecideaceae. In 1968, for example, Weber reduced 75 species names to two in one group of *Acarospora*.

In their usage of infraspecific ranks such as subspecies, variety and form, lichenologists endeavour to apply the principles and procedures of systematics that are used in other botanical groups, especially the vascular plants.

CHAPTER TWO

THALLUS STRUCTURE

2.1 Morphology

Mycobionts and photobionts of lichens can associate in a variety of ways, to form several different morphological types or growth forms. These can be interpreted as methods the mycobiont has evolved to display the photobiont in a manner designed to capture maximum irradiation and so ensure optimal photosynthesis.

The structurally simplest type of organization involves fungal hyphae enveloping single or small clusters of photobiont cells, and without any overall fungal layer. The resultant thallus, when growing superficially on rock or bark, often has a powdery appearance, and is referred to as *leprose*. This type of growth is seen in both evolving associations, as some Helotiales, and through to always sterile crusts as in *Lepraria* (figure 2.1*A*), a genus which must be viewed as highly derived because of its complex chemistry.

Structurally comparable lichens can be immersed beneath the surfaces of bark (*endophloeodic*; e.g. *Arthopyrenia fallax*, see figure 6.1*C*) or rock (*endolithic*; e.g. *Staurothele caesia*) but as these often involve some amount of layering they are referred to the crustose category. The outer layer may be only the cuticle in some *foliicolous* (leaf-inhabiting) lichens, such as *Strigula elegans* (figure 2.1*C*).

In *crustose* (crustaceous; figure 2.1*B*, *D–E*) types the photobiont cells are covered by a distinct layer of fungal tissue (*cortex*), or a mixture of hyphae with bark, epidermal cells or rock crystals. A consequence of the cortical layer is that the surface appears less powdery than in leprose types and is often smooth. Sometimes a distinct layer of loose fungal tissue is present below the photobiont layer; this, the *medulla*, is however often much less conspicuously developed than in foliose types. The surface of crustose thalli can be continuous, be dissected by wandering cracks (figure 2.1*E*), or comprise well-marked polygonal *areolae* as in *Rhizocarpon*. Areolae are adaptations to alternate wetting and drying regimes; when thalli are moist the tissues swell and the cracks close. Areolae sometimes grow out to form

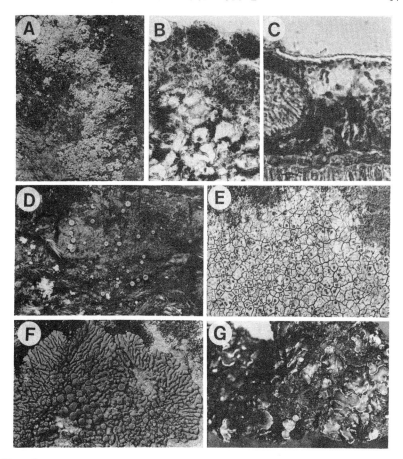

Figure 2.1 Leprose, crustose, placodioid and squamulose lichens. *A, Lepraria incana* (leprose), × 1.5. *B, Buellia punctata* (crustose), vertical section showing photobiont cells below a mycobiont cortex, with bark cells beneath), × 125. *C, Strigula elegans* (crustose, photobiont beneath the cuticle of the host leaf), × 30. *D, Dimerella diluta* (crustose, granular thallus), × 7. *E, Diploschistes caesioplumbeus* (crustose, cracked thallus), × 2.5. *F, Caloplaca thallincola* (placodioid thallus), × 2.5. *G, Psora decipiens* (squamulose thallus), × 1.5. Photographs *A* and *D–G* by F.S. Dobson.

coralloid tufted cushions and are then said to be *suffruticose*. Between areolae and at the growing margins of crustose thalli an algal free white or dark brown to black zone, the *prothallus*, containing only the mycobiont, is sometimes visible.

A variation on the crustose type, seen particularly in some groups of

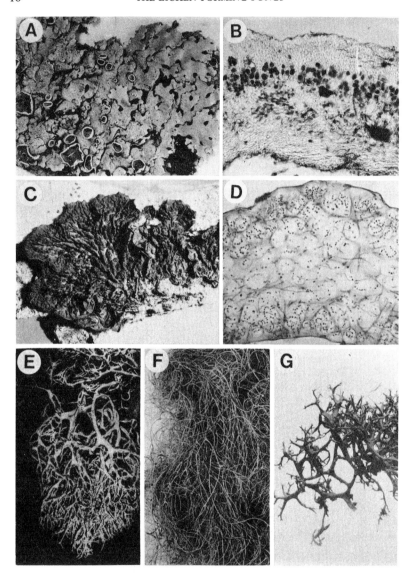

Figure 2.2   Foliose and fruticose lichens. *A, Parmelia quercina* (foliose), × 2. *B*, Vertical
section of *Physcia aipolia* (foliose), heteromerous thallus showing upper cortex, photobiont
layer, medulla and lower cortex, × 250. *C, Collema furfuraceum* (foliose), × 2.5. *D*, Vertical
section of *C. auriculatum,* homoiomerous thallus, × 300. *E, Letharia vulpina* (fruticose), × 1.
F, Bryoria subcana (fruticose), × 1. *G. Coelocaulon aculeatum* (fruticose), × 2.5. Photographs
A, C and *F–G* by F.S. Dobson.

Caloplaca and *Lecanora*, is the *placodioid* (placoid) in which the surface is radially striate with the marginal tissues slightly raised (figure 2.1*F*). Similar anatomically is the *squamulose* type but instead of the radially striate rosettes separate or overlapping scale-like squamules are formed; this is seen for example in *Catapyrenium*, *Coriscium* and *Psora* (figure 2.1*G*).

The layering of tissues is developed further in the foliose (leafy) types (figure 2.2*B*). In addition to the presence of a separate cortex, photobiont layer, and a usually proportionately very thick medulla, foliose lichens also have a clearly differentiated *lower cortex* which gives rise to special attachment organs (see section 2.4). The development of the lower tissues enables foliose thalli to be peeled from their substratum much more readily than is possible with crustose or placodioid growth forms. While most foliose lichens do have the layered structure described with the photobiont in a discrete layer (*heteromerous*; stratified), in a few genera such as *Collema* (figure 2.2*C, D*) the photobiont cells are dispersed throughout the thickness of the thallus (*homoiomerous*; unstratified).

The most conspicuous lichen thalli, however, belong to the *fruticose* (shrubby) type. These have a heteromerous structure which is not dorsiventral but developed around a vertical axis. The same type of structure is seen in both pendulous (figure 2.2*E, F*) species and erect tufted species (figure 2.2*G*). The supporting tissue can be either a thickened outer cortex, as in *Alectoria* and *Coelocaulon*, or a special central elastic axis (*chondroid axis*) in *Usnea* (figure 2.3).

All *Cladonia* species have two distinct growth-habits and they have sometimes been regarded as composite. The first-formed thallus structure

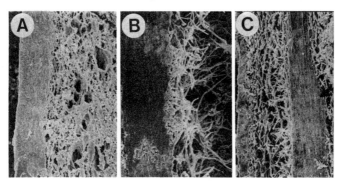

Figure 2.3 Structure of fruticose thalli. *A, Coelocaulon divergens*, × 70. *B, Alectoria ochroleuca*, × 75. *C, Usnea cavernosa*, note the central chondroid axis (arrow), × 75. Longitudinal sections, the cortex to the left. Scanning electron micrographs.

may be granular and short-lived, or squamulose and persistent depending on the species (figure 2.4). It is from these structures that fruticose growths develop but on the basis of developmental studies it is clear that these, termed *podetia*, are most correctly interpreted as lichenized apothecial stipes. Podetia may be richly branched (figure 2.4*D*) or cup-shaped *scyphi* (figure 2.4*C*, 3.5*B*) and bear pycnidia or hymenial tissues at their tips or on their rims respectively.

In all the cases mentioned above, it is the mycobiont which appears to have the predominant structural role. This is clearly not the situation in a few genera which have filamentous photobionts where the minute fruticose or threadlike thalli are produced by fungal hyphae sheathing the filaments (figure 2.5). These thalli, termed *filamentous*, are seen, for example, in *Coenogonium*, *Cystocoleus*, *Racodium* and some members of the Ephebaceae.

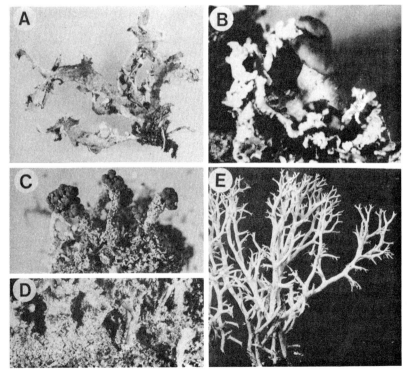

Figure 2.4 Variation in thallus types in *Cladonia*. *A, C. convoluta* (squamules dominant), × 1. *B, C. caespiticia* (squamules dominant with non-lichenized apothecial stipe), × 5. *C, C. floerkeana* (squamules and erect podetia with hymenial discs), × 2. *D, C. pocillum* (squamules and goblet-like podetia, 'scyphi'), × 1.5. *E, C. rangiformis* (squamules not persistent, fruticose podetia), × 2. Photographs *C* and *E* by F.S. Dobson.

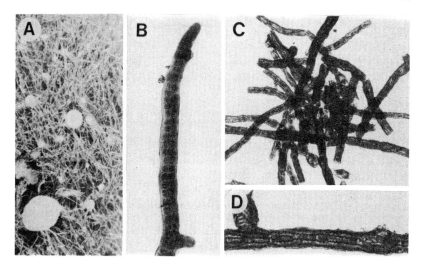

Figure 2.5 Filamentous lichens. *A, Coenogonium implexum*, × 12. *B, Ephebe lanata*, × 125. *C-D, Cystocoleus ebeneus*; *C* × 125, *D* × 135.

Foliose and fruticose lichen thalli are sometimes referred to as *macrolichens*, and crustose and other smaller types as *microlichens*.

In some cases the same growth-form will be developed by all the species in a genus or family, but this is not invariably so, for example in the Teloschistaceae (table 2.1) which includes a range of thallus types. In *Caloplaca* attempts to segregate genera based only on the habit, for example *Gasparrinia* for placodioid thalli and *Polycaulionia* for minutely shrubby suffruticose thalli have proved unsatisfactory owing to the presence of intermediate species.

Table 2.1 Growth forms developed in genera of the Teloschistaceae.

Genus	Growth form(s)
Leproplaca	leprose
Apatoplaca	crustose
Caloplaca	crustose, placodioid or suffruticose
Ioplaca	squamulose
Fulgensia	squamulose
Xanthopeltis	foliose (umbilicate)
Xanthoria	foliose
Teloschistes	fruticose

2.2   Environmental modifications

The environment can have profound influences on thallus form. This is
often particularly striking in the macrolichens (figure 2.6), but is also
widespread in crustose taxa where it has sometimes led to the introduction of
numerous unnecessary names (as in *Acarospora*; see section 1.7). The
availability of moisture, degree of exposure, substratum and abrasion
appear to be of particular importance. In some cases ball-like growths able
to be blown about by the wind may be developed, as in species of
Sphaerothallia (the 'manna' lichens) and several of *Parmelia* (e.g. *P.
revoluta*).

The environment can even result in the formation of distinct growth
forms within a single species. The usually crustose *Aspicilia calcarea* when
growing from rock on to adjacent soil can assume a fruticose habit in the
part of the thallus on the soil. Conversely, the normally fruticose
Pseudephebe miniscula can become almost crustose in the central part of its
rosettes in extreme arctic and antarctic situations. When growing under
particularly shaded and humid conditions the normally crustose
Baeomyces rufus can develop minute squamulose structures.

Other characters subject to environmental modifications include the
distance between areolae in crustose species, a frosted appearance or
pruinosity (see section 6.4), and the production and size of ascomata and
vegetative propagules.

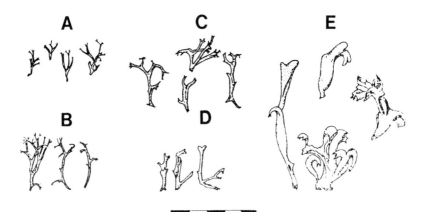

Figure 2.6   Specimens of *Cladonia uncialis* subsp. *biuncialis* from various habitats in the
British Isles. *A*, Dry heath. *B*, Sand dunes below pines. *C*, Shallow peat. *D*, Shingle by river. *E*,
Very wet peat. Scale = 5 cm. From D.L. Hawksworth (1973), in *Taxonomy and Ecology* (V.H.
Heywood, ed.), Academic Press, London and New York, 31–69.

2.3 Tissue types

Fungal hyphae can give rise to a range of tissue types and these have been increasingly recognized as of importance in the systematics of non-lichenized fungi. There is almost no information on tissue types in crustose lichens, apart from those in association with ascomata, but in several foliose and fruticose groups they have proved to be of major importance. Their study has been greatly facilitated by the use of the scanning electron microscope (SEM) as this renders the generally colourless, thick, and often fused wall layers of the hyphae opaque.

The *cortex* in macrolichens normally consists of a single tissue type and two main types can be distinguished, *paraplectenchyma* where the cells are randomly orientated giving a cellular appearance, and *prosoplectenchyma* where elongate hyphae are orientated in a particular direction. However, in *Ramalina* a paraplectenchymatous layer is overlain by a prosoplectenchymatous one, and other variations are known. In 1973 Hale, using the SEM, distinguished a thin polysaccharide layer overlying the cortex in some groups of Parmeliaceae (especially *Parmotrema*) and

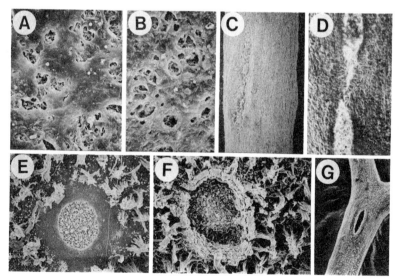

Figure 2.7 Pored epicortex, pseudocyphellae and cyphellae. *A, Parmelia conspersa* epicortex, × 275. *B, Parmeliopsis aleurites* epicortex, × 275. *C, Alectoria ochroleuca* pseudocyphella, × 25. *D, A. nigricans* pseudocyphellae, × 300. *E, Sticta sylvatica* cyphella, × 75. *F, S. weigelii* cyphella, × 55. *G, Oropogon loxensis* cyphella-like pore, × 30. Scanning electron micrographs.

other families; this *epicortex* often has pores of 10–20 μm diam. which Hale suggests may be important for gaseous exchange (figure 2.7*A,B*).

In contrast to the compact cortical tissues, the *medulla* is almost always composed of loosely interwoven hyphae with clear gaps between them (figure 2.2*B*), and may be termed *chalaroplectenchyma*. Crystals of lichen products can often be seen to encrust the surfaces of medullary hyphae when viewed with the SEM.

2.4   Attachment organs

There is a surprising variety in the organs developed to attach lichens to their substratum (figure 2.8). In the case of species lacking a lower cortex, this is achieved by hyphae from the medullary tissues growing into the surface layers of the substratum, passing between cells or crystals of minerals (p. 84). In some instances, particularly on soils, certain crustose and squamulose species can develop complex irregularly branched tough *rhizinose strands* (e.g. *Squamarina gypsaceus*) or a delicate reticulately branched *hyphal net* (e.g. *Psora decipiens*) binding the soil particles together.

In foliose lichens, attachment is normally by *rhizines* (figure 6.1*B*), which can be simple or branched in a variety of ways. Different types of rhizines prove valuable for generic and species separations in some cases, especially in the Parmeliaceae. *Anzia* is notable in having a very thick black reticulate spongy photobiont-free tissue on the lower surface, a *hypothallus*. In *Hypogymnia*, however, rhizines are absent and the attachment is by hyphae penetrating the substratum. A similar method is used in some groups of *Ramalina* and *Usnea* where the base can be a blackened and persistent algal-free *holdfast*, which may penetrate into the substratum (figure 6.1*A*). Foliose thalli attached at a central point, as in *Lasallia* and *Umbilicaria*, are termed *umbilicate*.

Large pendulous or straggling species can lose the attachment to their point of origin and drape over vegetation, occasionally (e.g. *Alectoria sarmentosa* subsp. *vexillifera*) fixed by *hapters*, short branches the apices of which penetrate the outer tissues of the branches of ericaceous shrubs.

2.5   Cyphellae and pseudocyphellae

Neat circular structures, *cyphellae*, are a characteristic feature of the lower surfaces of *Sticta* species (figure 2.7*E,F*). These open into a depression lined by somewhat specialized rounded cells and have a role in facilitating

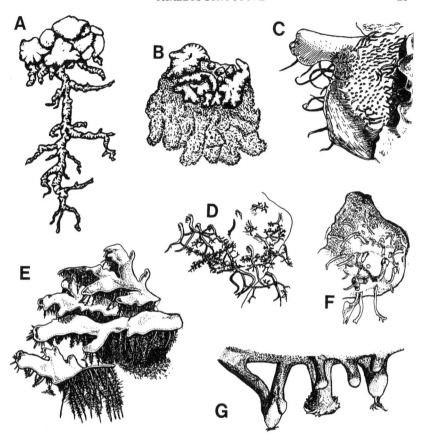

Figure 2.8   Attachment organs in squamulose (*A-B*) and foliose (*C-G*) lichens. *A, Buellia pulchella* rhizinose strand, × 8. *B, Psora decipiens* hyphal net, × 8. *C-F*, Rhizines. *C, Parmelia arnoldii*, × 8, *D, P. taylorensis*, × 15. *E, Physconia pulverulacea*, × 6. *F, Parmelia rudecta*, × 15. *G, P. septentrionalis*, × 30. Adapted from B. Hannemann (1973), *Bibliotheca lich*. **1**, 1–123.

gaseous exchange (section 5.4). Similar but less differentiated structures on the thallus of certain *Pyrenula* species, and distinct holes through to the medullary cavity in *Oropogon*, are often also referred to as cyphellae.

The same function is attributed to *pseudocyphellae* in which looser hyphal medullary tissue comes to the surface in discrete patches from which the cortex is absent (figure 2.7*C, D*). These have the same lichen products as the medullary tissue which gives rise to them and so may give different reagent tests to the cortex surrounding them. Pseudocyphellae are much more widespread than cyphellae occurring, for example, in *Alectoria, Bryoria, Cetrelia, Coelocaulon, Pseudocyphellaria*, and *Punctelia*. Cyphellae and pseudocyphellae do not occur in lichens with a pored epicortex, a structure which may have an identical role (see section 2.3).

2.6 Cephalodia and the role of the bionts in morphogenesis

In the case of three-membered symbioses involving two photobionts, a green alga generally predominates in the thallus and cells of a blue-green cyanobacterium are restricted to specific parts or structures, *cephalodia* (figure 2.9). Cephalodia occur in about 520 species distributed through some 21 genera in eight families and may be divided into two main types: (1) *internal cephalodia* confined to particular parts of the thallus, such as between the 'ribs' of *Lobaria pulmonaria*, or forming a continuous or discontinuous layer below the green photobiont as is the case in *Solorina crocea*; and (2) as *external cephalodia* where morphologically distinct structures are developed. The latter can appear as superficial wart-like outgrowths as in *Peltigera aphthosa* and *Placopsis gelida*, or minutely fruticose outgrowths as in *Lobaria amplissima*.

In all cases, it is presumed that the cephalodia arise from blue-green photobiont cells, alighting by chance on the thallus, being enveloped by the mycobiont, and taken into its structure. As cyanobacteria are able to fix atmospheric nitrogen (see section 5.6), their capture and growth could be expected to be advantageous to the thallus. Only a single cyanobacterium is usually involved but *Nephroma arcticum* has been found to have two types (in separate cephalodia) occasionally on a single thallus.

Since Dughi's pioneering studies in 1937–45 it has been recognized that the fruticose cephalodia of *Lobaria amplissima* live independently from their lichen host; indeed these and other detached fruticose cephalodia have been referred to a separate genus, *Dendriscocaulon*. However, it was only with the investigations of James and Henssen in the 1970s that the morphogenetic significance of cephalodia started to be appreciated—a single mycobiont developing totally different growth forms depending on whether the photobiont was green or blue-green. Further, in some cases the lichen products produced with a green photobiont were not formed with the blue-green partner.

Several particularly spectacular cases have now been documented, especially in *Stica* (figure 2.9*D*) and *Peltigera*, in a few instances also involving green algae captured by a blue-green photobiont-containing

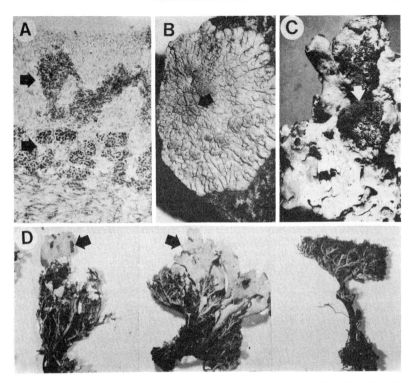

Figure 2.9 Types of cephalodia in lichens. *A, Solorina crocea*, vertical section of thallus showing (arrows) the layering of blue-green (internal cephalodia) and green photobionts, × 325. *B, Placopsis gelida*, external cephalodia (arrow) arising in the thallus centre, × 3. *C, Lobaria amplissima*, external minutely fruticose cephalodia (arrow) erumpent from the thallus surface, × 2. *D, Sticta felix* green photobiont-containing foliose lobes (arrows) arising from a blue-green photobiont-containing fruticose thallus (right), × 1.

thallus. In both these genera there are examples of joined thalli which can also be found independently and have been given separate species names—such thalli have been termed 'chimeroid', 'phycotypes' or 'symbiodemes'.

Occasionally, cyanobacterial cells can be found scattered in or on the cortex but not in discrete cephalodia, as in *Roccellina niponica*.

2.7 Culture and synthesis

When grown in pure culture, lichen mycobionts and photobionts tend to produce rather structureless colonies which have little similarity in their

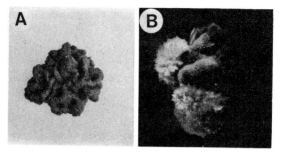

Figure 2.10 Isolated bionts in pure culture. *A, Pseudotrebouxia impressa* from *Physcia stellaris*, × 1. *B, Stereocaulon dactylophyllum* mycobiont, × 1.5. Photographs by V. Ahmadjian.

appearance to the thallus from which they were derived (figure 2.10). This, together with the recognition of the morphogenetic implications of cephalodia, leaves no doubt that the development of a lichen thallus is a result of the interplay between the partners involved.

During the last 25 years, Ahmadjian and his co-workers have made tremendous progress in culturing lichen bionts and endeavouring to re-combine them. In 1970 the successful re-synthesis of *Endocarpon pusillum*, first reported by Stahl in 1877, was repeated in full. This was a turning point and since that time their successes have continued to increase. A key feature in this has been the appreciation of the need for wetting-drying regimes and the transferring of bionts to sterile soil after initial contact had been made on agar plates. Their work with *Cladonia cristatella* has been spectacular; not only have squamules, podetia, hymenium, pycnidia and lichen products been produced within about five months, but it has been possible to conduct synthesis experiments with different biont isolates. These have already contributed significantly to our knowledge of photobiont speci-ficity (section 1.6) and lichen product biosynthesis (section 8.1).

Macerated lichen thalli have also now been spread on sterile soil and grown on into typical thalli under controlled 'phytotron' conditions. This technique can be expected to be of value in many areas of experimental lichenology in the future.

CHAPTER THREE

REPRODUCTION

This chapter describes the various types of reproductive structures found in the lichen-forming fungi, and reviews their development and roles. Dispersal and establishment are considered separately in Chapter 4.

3.1 Ascomata

Ascus-bearing structures, the preferred general term for which is *ascomata* (sing. ascoma), differ substantially in the different orders of fungi which include lichenized representatives, and to a lesser extent in families within those orders. This is to be expected as ascomatal characters are of major importance in the delimitation of higher taxonomic ranks within the Ascomycotina. These can be divided into two main morphological types on the basis of their superficial appearance: *apothecia* in which the ascus-bearing tissue (*hymenium*) is exposed at maturity (figure 3.1), and *perithecia* in which the asci are borne within a locule (figure 3.2). It is essential to remember that these types are end products and that their development (*ontogeny*) can be attained in distinct ways. More complex terminologies endeavouring to allow for ontogenetic differences have been proposed but are of limited value and not recommended here.

Apothecia may be weakly delimited patches, as in the tropical *Cryptothecia* (Arthoniales), but have well-defined limits in most other groups. The hymenium is most commonly disc-like and sessile, but in some instances can have a distinct stalk (*stipe*) lacking photobiont cells which may be colourless (e.g. *Baeomyces rufus, Cladonia caespiticia*) or pigmented (Caliciales); the variety of types is illustrated in figure 3.1. In the case of most *Cladonia* species, the stipes are lichenized and termed *podetia* (figure 2.4). Elongate apothecia exposing the disc by a narrow or broad slit, *lirellae*, arise in a variety of orders; they are particularly characteristic of some genera in the Graphidales and Opegraphales (figure 3.1*B*).

The tissue supporting the margins of the discs in apothecia (the excipulum) may lack photobiont cells (excipulum proprium, proper margin, *true exciple*) when its colour is usually similar to the disc or stalk. Alternatively, it may be combined with an outer photobiont-containing

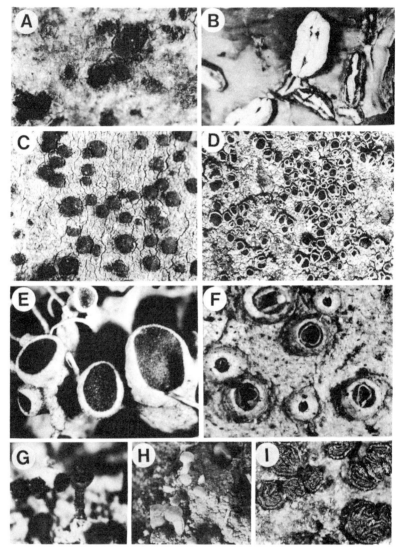

Figure 3.1 Examples of lichen apothecia. *A, Arthonia tumidula* (weakly delimited, 'ardellae'), × 10. *B, Graphis afzelii* (lirellae), × 10. *C, Lecidella elaeochroma* (lecideine apothecia), × 6. *D, Rinodina roboris* (lecanorine apothecia), × 4. *E, Physcia semipinnata* (lecanorine apothecia), × 3. *F, Ocellularia americana* (note separate thalline and true exciples), × 8. *G, Calicium viride* (stipitate mazaedium), × 15. *H, Baeomyces roseus* (stipitate apothecium), × 5. *I, Umbilicaria torrefacta* (gyrose apothecium), × 5. Photographs *D* and *H* by F.S. Dobson.

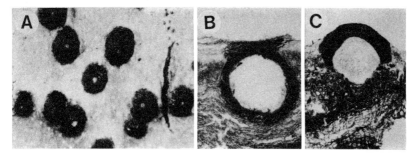

Figure 3.2   Examples of lichen perithecia. *A, Pyrenula chlorospila* surface view, note white ostioles, × 15. *B, P. nitidella,* vertical section of entire perithecium, × 50. *C, Acrocordia gemmata,* vertical section of dimidiate perithecium, × 50.

tissue (excipulum thallinum, thalline margin, *thalline exciple*) whose internal structure recalls that of the thallus; in these cases the margin is almost invariably concolorous with the thallus. Apothecia with only a true exciple are *lecideine* (figure 3.1*C*) , while those also with a thalline exciple are *lecanorine* (figure 3.1*D-E*); the lecanorine type is of course restricted to lichenized groups, but some lichenized genera include species with lecanorine and lecideine margins (e.g. *Caloplaca*). In *Ocellularia* (figure 3.1*F*) and *Thelotrema* the two exciple types separate so that from above the apothecia appear to have one margin within another.

The ascus-containing layer (*hymenium*) arises from a generative layer (*subhymenium*) which may be supported by a *hypothecium* layer; paraphysis tips may form a distinct layer (*epithecium*) above the ascal layer.

Perithecia usually arise singly, but in a few genera are combined into a common stroma as in the largely tropical genus *Melanotheca*. The wall surrounding the generative locule (*centrum*) is referred to as the *exciple* or *peridium*; the wall may be continuous below the locule (*entire*) or cone-like and not extended below the locule (*dimidiate*) as shown in figure 3.2. The type of tissue from which the exciple is made may be cellular, hyphal, or a mixture of bark cells and hyphae (a *clypeus*). Asci in perithecia discharge by way of a pore, usually apical, the *ostiole*. In some genera a separate carbonized shield-like layer (*involucrellum*) extends outwards from the ostiole, as in some *Verrucaria* species; when an involucrellum is present the exciple is usually relatively well-developed and often almost colourless.

In species of a few genera, for example *Endocarpon, Gonohymenia, Staurothele* and *Thelenidia*, photobiont cells occur in the ascomatal cavity or amongst the asci (see section 4.3).

3.2 Ontogeny

In the non-lichenized Ascomycotina, only two main developmental types were thought to occur until the 1960s: (1) *ascohymenial* where asci arise in a specially formed hymenium, and (2) *ascolocular* in which they arise from cavities in a pre-formed stroma. The ascohymenial type was considered to be exclusively associated with unitunicate asci and the ascolocular with bitunicate asci. Such an interpretation has proved to be unacceptably simplistic—there are many more ascus types (see section 3.4) and studies on mainly lichenized orders, especially by Henssen in Germany and Letrouit-Galinou in France, have revealed considerable variations in ontogeny and several distinctive types. These latter include ascomata developing from pycnidia in *Ephebe* and allied genera (*pycnoascocarps*), asci arising amongst thalline hyphae in gall-like swellings (*thallinocarps*) in *Gonohymenia* etc., clearly ascohymenial types with bitunicately discharging asci as in *Peltigera*, and ones intermediate between ascolocular and ascohymenial types in the Arthoniales and Opegraphales.

The hymenium may be exposed from the earliest stages (*gymnocarpic* as in most Lecanorales, see figure 3.3), enclosed at first but opening before fully mature (*hemiangiocarpic* as in Graphidales, Peltigerales and Pertusariales), or remaining closed at least until the spores are mature (*angiocarpic*, as in all genera with perithecia).

Space does not permit a discussion of the various ontogenetic types here; the comprehensive well-illustrated survey included in the textbook of Henssen and Jahns (see Further Reading) should be consulted for further information on this aspect. However, it is of interest to note that in many lichenized groups the onset of ascomatal development involves the production of a group of specialized cells which give rise to erect filaments projecting above the level of the cortex. These are the *trichogynes* which are able to receive conidia produced by pycnidia of the same lichen and through which the nuclear contents of the conidium are transferred to the ascomatal initial within the thallus (see section 3.2, figure 3.8). Conidia with a sexual function are referred to as *spermatia*. The available evidence suggests that the process is similar to that seen in non-lichenized fungi, i.e. that the nucleus from the fertilizing conidium is used to establish a heterokaryon and that the two nuclear types conjugate and undergo genetic recombination by meiosis in the asci prior to dividing mitotically to produce the ascospores. The cytological processes involved are difficult to observe in most lichens (as in other fungi) as the nuclei and chromosomes are minute.

The method by which a heterokaryon is produced in those lichens which are not known to form trichogynes or pycnidia with spermatia is uncertain,

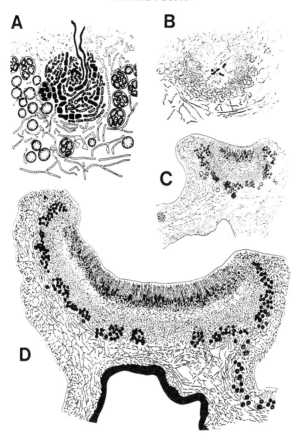

Figure 3.3 Development of the apothecium in *Parmelia exasperata*. *A*, Generative tissue with ascogonia. *B*, Primordium. *C*, Young developing apothecium. *D*, Mature apothecium. From A. Henssen and H.M. Jahns (1973), *Lichenes*, Thieme, Stuttgart.

but the necessity for such a complex system has been avoided in some non-lichenized fungi, for example by the parasexual cycle.

3.3   Hamathecia

The nature of the *hamathecium*, a general term introduced in 1981 for tissues separating asci in all ascomycetes, is dependent on the ontogeny of the ascomata. Their study has consequently assumed great importance, although interascal tissues were almost entirely ignored until the last 10–15 years.

The terminology of the different types is based on their place of origin and direction of growth (figure 3.4). *Paraphyses* originate from the base of

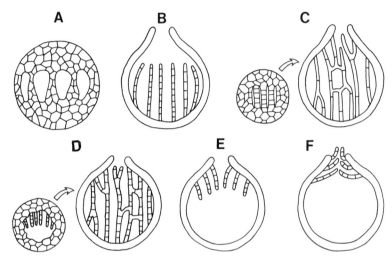

Figure 3.4 Hamathecium types. *A*, Interascal pseudoparenchyma. *B*, Paraphyses. *C*, Paraphysoides (trabecular). *D*, Pseudoparaphyses (cellular). *E*, Periphysoids. *F*, Periphyses. *G*, Absent (not illustrated). Adapted from O. Eriksson (1981), *Opera bot.* **60**, 1–240.

the ascoma and grow upwards; *paraphysoids* from the stretching of tissues present before the asci develop; *pseudoparaphyses* from above the level of the asci growing downwards and finally becoming attached to the base; *periphysoids* from above the asci and growing down a short distance; and *periphyses* which line the ostiolar canal usually orientated upwards. In some groups such as the Verrucariaceae, no hamathecial elements are present at maturity, and in a few it comprises largely unchanged cellular tissues present when ascus formation started.

The recognition of paraphysoids, which are usually thin, sparsely septate, and with net-like anastomosing branches, has been a key factor in the segregation of genera, for example of *Huilia* from *Lecidea*, and of *Micarea* from *Catillaria*. In *Catillaria* and *Lecidea* true paraphyses occur; these are thick, regularly septate, usually unbranched, not anastomosed, and with swollen pigmented apices.

3.4 Asci

Studies over the last two decades have shown that asci belong to several fundamentally different types. The transmission electron microscope (TEM) has been a key to the elucidation of their structure as light microscopical studies had often proved difficult to interpret. Ascus walls

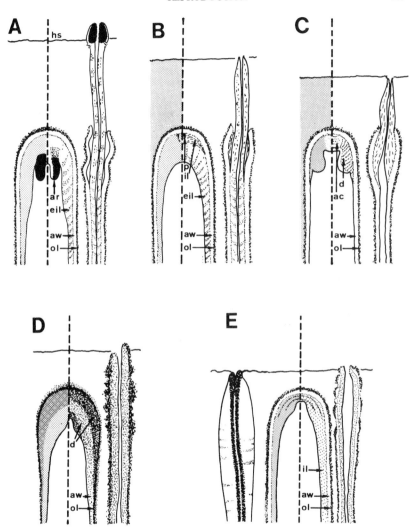

Figure 3.5 Examples of ascus types in lichen-forming fungi as seen by light microscopy (left) and transmission electron microscopy (right). *A, Peltigera*-type. *B, Rhizocarpon*-type. *C, Lecanora*-type. *D, Teloschistes*-type. *E, Pertusaria*-type. *A-B*, fissitunicate; *C-E*, non-fissitunicate. Dark zones on left = amyloid reactions; dark zones on right = electron-dense. *ac* apical cushion; *ar* eversible amyloid ring; *aw* ascus wall; *d* apical dome; *eil* expansible inner layer; *hs* hymenium surface; *il* inner layer; *ol* outer layer. Adapted from R. Honegger (1982), *J. Hattori bot. Lab.* **52**, 417–429.

are now seen to be invariably multi-layered, and consequently terms such as *unitunicate* and *bitunicate* (implying one and two wall layers respectively) which have come to be widely used since the 1950s are misleading. It is more appropriate to refer to asci in which the inner wall layers (*endotunica*) separate from the outer layers (*ectotunica*) and elongate substantially prior to spore release and act in a 'jack-in-the-box' manner as *fissitunicate*, and ones where there is no such extension and 'jack-in-the-box' action as *non-fissitunicate*. Fissitunicate types usually have a finger-like projection (*ocular chamber*) extending into a thickened apex (*tholus*); and the extruded endotunica reaches to the ostiole in perithecioid genera before it discharges its spores. In most families of the Lecanorales so far studied, at discharge the multi-layered walls undergo a more limited extension (usually just to the hymenium surface); in *Porina* s.str., where discharge is through an apical pore, Eriksson suggests that the inner layers which extend in fissitunicate types have been reduced to an apical ring! It must therefore be stressed that these terms are descriptive of function rather than structure.

Positive (*amyloid*) reactions with iodine are of particular interest in studies of asci. For example, asci can be completely unaffected (e.g. *Arthopyrenia, Pyrenula*), a large part of the dome-like apical thickening may stain (e.g. *Lecanora, Parmelia*), or the reaction may be restricted to a thin outer, and very thin inner, layers of the whole ascus (e.g. *Pertusaria*).

In the Caliciaceae the non-amyloid asci disintegrate as the spores mature leaving these in a dry powdery mass (*mazaedium*) well-suited for dispersal by wind or insects crawling over them.

Some of the types of asci which have already been recognized amongst the lichen-forming Ascomycotina are shown in figure 3.5, but it should be remembered that many have yet to be studied critically. It is of interest to reflect that the types seen in the Lecanorales and Teloschistales could be theoretically derived from asci with the structure of those in Peltigerales—as could the less complex fissitunicate types of the largely non-lichenized orders Dothideales and Pyrenulales (by the loss of the iodine-positive tissues) or the non-lichenized Helotiales and Sphaeriales (by the loss of the outer wall layers).

The physiology of ascus discharge has not been critically examined in lichenized fungi but there is no reason to suspect that the processes differ from those of other ascomycetes.

3.5 Ascospores

Ascospores in lichen-forming fungi vary greatly in size, shape, structure and septation and may be colourless, greenish or brown (figure 3.6). They

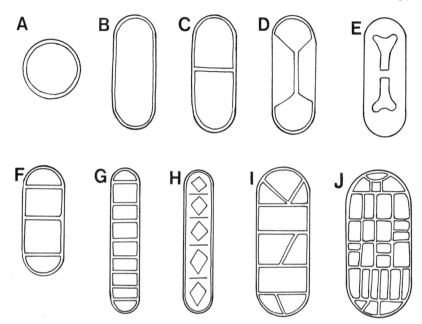

Figure 3.6 Examples of ascospore types in lichen-forming fungi. *A*, Globose and simple (1-celled). *B*, Ellipsoid and simple (1-celled). *C*, 1-septate (2-celled). *D*, Polarilocular. *E*, Mischoblastiomorph. *F*, 3-septate (4-celled). *G*, 8-septate (9-celled). *H*, 4-distoseptate (*Pyrenula*-type). *I*, Submuriform. *J*, Muriform.

may lack all septa and be *simple*, have transverse septa alone, or have a mixture of transverse and longitudinal septa, when they are termed *muriform*. The septa themselves may be continuous with the outer spore wall (*eusepta*) or be developed secondarily within the original wall (*distosepta*) as in *Pyrenula*. In most members of the Teloschistales the septum is thickened and penetrated by a thin canal; these spores are referred to as *polarilocular*. Some particularly complex variations on the polarilocular theme occur in the Physciaceae, especially *Rinodina*. In *Graphina*, *Thelotrema* and allied genera of the Graphidales the ascospores usually give a violet reaction with iodine.

Spore ornamentation is pronounced in many members of the Caliciales but is generally rare in lichenized taxa even when these are studied with the SEM. Gelatinous sheaths, often swelling visibly in potassium hydroxide, are conspicuous in a few genera (e.g. *Lichenothelia*).

Eight ascospores are usually produced in each ascus, but the numbers can vary from one in certain *Mycoblastus* species to several hundred in some of *Acarospora*. Both extremes may confer some advantage; large multinucleate thick-walled spores would be expected to survive for long periods, while increased numbers would increase the probability of making contact with an appropriate photobiont.

3.6 Conidiomata

Specialized multi-hyphal structures which bear conidia are termed *conidiomata* (sing. conidioma). These are widespread in the lichen-forming Ascomycotina, although after the pioneering studies by Tulasne and Lindsay in the 1850s and 1860s they were scarcely studied critically until the late 1970s! They are still inadequately studied and receive scant treatment in most texts but their taxonomic significance both at the species level and at higher ranks is now starting to be recognized. The condiomata found in lichen thalli are almost invariably immersed, flask-shaped or subglobose *pycnidia*. An exception is afforded in *Micarea adnata* where the conidia can arise over the surface of a compact tissue to form a *sporodochium*. The peculiar stalked peltate *hypophores* of some genera of tropical leaf-inhabiting Asterothyriaceae (e.g. *Echinoplaca, Tricharia*) should perhaps also be interpreted as conidiomata.

At least five types of lichen pycnidia can be recognized on developmental grounds (figure 3.7) and eight ways in which the cells or hyphae (*conidiophores*) supporting the conidiogenous cells can be arranged are documented. Surprisingly, in all the methods by which the conidia are produced, the process of *conidiogenesis*, appears to be phialidic (i.e. the conidia are not formed from existing wall-layers of the conidiogenous cells but by newly delimited tissue). This is so whether the conidia are produced terminally, laterally or in an intercalary manner. Percurrent proliferation is particularly marked in *Lecanactis*, where annellations are clearly visible; it is less distinct in other genera studied where conidia can form at almost the same level.

The *conidia* (sing. conidium) themselves vary in shape and size but are most frequently cylindrical, sickle-shaped or filiform, and apparently always colourless. Most lichen species only produce one type of conidium, but many may have two, as in *Anisomeridium biforme* and *Strigula elegans*; one is then generally larger (*macroconidia*) than the other (*microconidia*); these are usually formed in different pycnidia but can be in the same one. Several *Umbilicaria* species in the *U. vellea* group develop superficial thick-walled dark-coloured simple or multi-cellular '*brood-grains*' on their undersides and rhizines in addition to pycnidia. In several *Micarea*

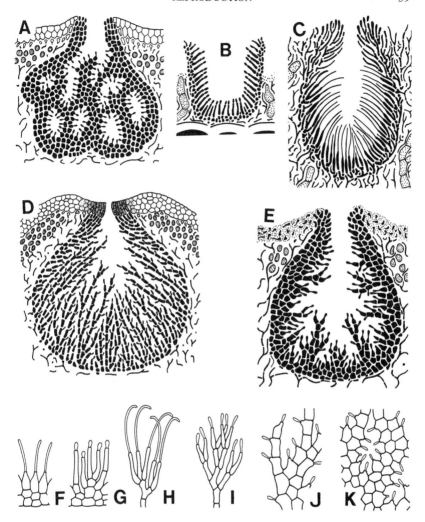

Figure 3.7 Examples of pycnidium (*A-E*) and conidiophore (*F-K*) types in lichens. *A, Xanthoria*-type. *B, Lecanactis*-type. *C, Roccella*-type. *D, Lobaria*-type. *E, Umbilicaria*-type. *F, Arthonia galactites. G, Lecanactis*- and *Peltigera*-type. *H, Roccella*-type. *I, Cladonia*- and *Ramalina*-type. *J, Anaptychia*- and *Lobaria*-type. *K, Dermatocarpon*- and *Xanthoria*-type. Adapted from G. Vobis and D.L. Hawksworth (1981), in *The Biology of Conidial Fungi* (G.T. Cole and B. Kendrick, eds.), 1, Academic Press, New York and London, 245–273.

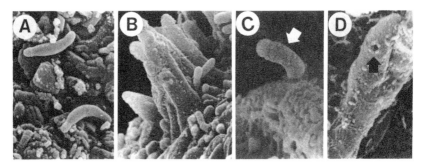

Figure 3.8 Conidia and trichogyne contacts in *Cladonia furcata*, scanning electron micrographs. *A*, Conidia, × 3000. *B*, Trichogynes, × 450. *C*, Conidium (arrow) attached to trichogyne, × 8500. *D*, Circular hole (arrow) in trichogyne where conidium was attached, × 5000. Adapted from R. Honegger (1984), *Lichenologist* **16**, 11–19.

species, for example *M. denigrata*, three separate types have been recognized—more pycnidial types than known in any other ascomycete species whether lichenized or not.

The role of conidia is uncertain and most probably varies from species to species. In some cases they are formed much more commonly than ascomata and function as propagules, as in *Cliostomum griffithii* and *Opegrapha vermicellifera*. In a few cases they are the only propagules known; in such instances lichens have been referred to ascomycete genera on other criteria (e.g. *Lecanactis subabietina*) but in others generic names in Coelomycetes or Hyphomycetes have been introduced (see section 1.5). However, in the majority of the Lecanorales it is more likely that their role is sexual (see section 3.2); the SEM has recently been used to demonstrate this process (figure 3.8). Such dual roles are not surprising as there are parallels in many groups of non-lichenized Ascomycotina, and where one lichen produces more than a single conidium type one may have a role in propagation and the other in fertilization.

3.7 Vegetative propagules

Ascospores and conidia disperse only the mycobiont of a lichen. However, special vegetative propagules able to disperse mycobiont hyphae and photobiont cells simultaneously have developed in many species. The commonest of these is the *soredium* (pl. soredia), a cluster of photobiont cells enveloped in a weft of hyphae (figure 3.9*C*). These may arise over the whole surface of a thallus, wherever the cortex is lacking, or in discrete

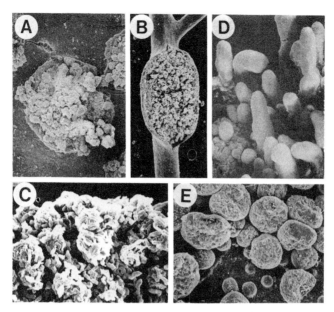

Figure 3.9 Soredia and isidia. *A, Parmelia sulcata* soralium, × 55. *B-C, Bryoria simplicior*; *B*, soralium, × 35; *C*, soredia, × 3500. *D, Pseudevernia furfuracea* coralloid isidia, × 35. *E, Parmelia pastillifera* globose isidia constricted basally, × 55. Scanning electron micrographs.

patches (*soralia*); their precise position and shape often provides valuable criteria for species separation. Patches of soredia are dry and powdery, rendering them difficult to wet, but generally adhere to fingers or material brushed against them.

In contrast to a soralium, an *isidium* (pl. isidia) is formed as an outgrowth from the cortex and so has a smooth appearance and discrete structure (figure 3.9 *D, E*). Photobiont cells are included within these papilla-, coral-, petal- or scale-like excrescences which tend to become constricted basally so that they are easily broken away from the thallus when touched. Soralia and isidia do not occur in the same lichen, although soredia starting to grow into filaments while still in soralia can sometimes give such an impression in some fruticose *Bryoria* and *Usnea* species. More substantial structures differing from isidia principally in size, the *fibrils* in species of *Usnea* (e.g. *U. longissima*) and *spinules* in some of *Bryoria* (e.g. *B. bicolor*) can also be narrowed basally and break away from the parent thallus.

More rarely occurring types of propagules (figure 3.10) include *phyllidia*, abstricted leaf- or scale-like dorsiventral portions of the whole thallus in

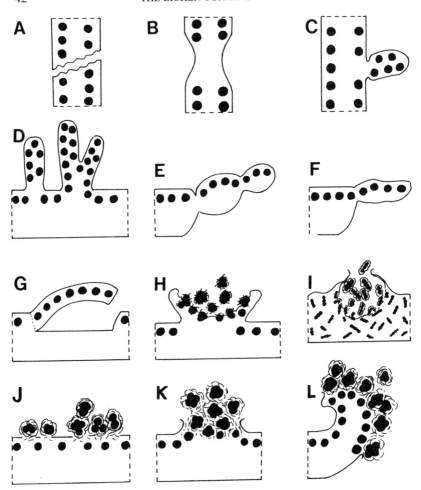

Figure 3.10 Types of vegetative propagules in lichens. *A*, Thallus fragmentation. *B*, Fragmentation region. *C*, Lateral spinule. *D*, Isidia. *E*, Blastidia. *F*, Phyllidium. *G*, Schizidium. *H*, Goniocyst. *I*, Hormocyst. *J*, Soredia formed from eroded surface. *K*, Soredia formed in soralium. *L*, Soredia formed from recurved lower cortex (labriform).

some foliose taxa (e.g. *Collema flaccidum, Peltigera praetextata*); *schizidia*, upper layers of the thallus splitting away as scale-like segments as in *Fulgensia bracteata* subsp. *deformis* and *Parmelia taylorensis*; *blastidia*, segmenting yeast-like thalline propagules in *Physcia opuntiella*; *goniocysts*, a photobiont cell and its offspring wrapped in hyphae so not soralium-like (formed in special *goniocystangia* in some tropical foliicolous lichens); and

the *hormocysts* of several *Lempholemma* species where photobiont filaments and hyphae grow together in a chain-like manner which break into
clumps—these last arise in special *hormocystangia*. Hypophores (see
section 3.6) sometimes also form dual propagules comprising photobiont
cells and mycobiont hyphae.

Probably a major but often unrecognized method of vegetative propagation is by thallus *fragmentation*. This is of especial importance in
fruticose lichens which are often fragile when dry and easily break when
trampled, as in *Cladonia stellaris* and *C. uncialis*. The resultant fragments
may be blown about or carried in fur or clothing; they can be conspicuous
in compacted drifted snow in the Arctic. In *Bryoria capillaris* small sections
of branches are narrow, devoid of algae, and either colourless or
blackened; these *fragmentation regions* are weak points easily broken by
twig movements or strong winds.

3.8 Evolutionary trends in reproductive strategies

Most of the non-lichenized Deuteromycotina (imperfect fungi) in which no
sexual state (teleomorph) is known, have been derived from groups within
the Ascomycotina. There is evidently substantial adaptive value in opting
for asexual reproductive methods in the group and this is reflected in the
lichen-forming representatives. The production of large numbers of
propagules comprising both bionts enables niches to be exploited rapidly.
The majority of widely distributed lichens rely largely on asexual methods
of propagation (e.g. *Chrysothrix candelaris*, *Evernia prunastri*, *Hypogymnia
physodes*, *Parmelia sulcata*, *Pseudevernia furfuracea*). The adoption of
purely asexual methods is, however, at the expense of an ability for genetic
recombination; such species are in evolutionary dead-ends.

Studies of the number of representatives of different genera which
produced isidia or soredia by Bowler and Rundel in 1975 made it clear that
crustose taxa tend to have few sorediate or isidiate representatives. They
noted, for example, that only one of 118 species of *Aspicilia* is sorediate,
and only 3% of foliicolous lichens had vegetative propagules. In contrast,
they are most abundant in fruticose families such as the Ramalinaceae and
Usneaceae; for example about 38% of the described *Usnea* species have
soredia or isidia.

In many cases species in which no ascomata are produced can be reliably
placed in genera where these are known on the basis of anatomical and
chemical characters; others form these but only rarely. Where ascomata
are produced at low frequencies, for example 0.4% of specimens of *Bryoria*

capillaris collected in North America, they are clearly not a major factor in propagation and may even be relict, failing to produce ascospores.

A few cases are known in which the method of propagation used varies with the age of the lichen. In *Peltigera spuria* only soredia are produced in young thalli, while in older ones only apothecia are present. Also of interest is *Lecidea verruca*, a lichenicolous lichen which appears to be dioecious; here pycnidia, producing conidia which are presumed to function as spermatia, occur on separate thalli that are also smaller than those in which trichogynes and later apothecia are formed.

Many cases are now known of *species pairs* (Artenpäare), one of which forms ascomata (*primary species*) and the other only asexual propagules (*secondary species*) but which are indistinguishable in all other respects; the secondary species is usually the most widely distributed.

Examples of some species pairs are given in table 3.1. In a few cases more than a single secondary species may have arisen from one primary species, as with *Parmelia reticulata* (sorediate) and *P. subisidiosa* (isidiate) from *P. cetrata*, and the two differently sorediate species *Physcia adscendens* and *P. tenella* from *P. semipinnata*. In many instances, however, primary species of now widespread lichens (e.g. *Pseudevernia furfuracea*) no longer appear to be extant but can be postulated to have once existed.

Table 3.1 Examples of species pairs

Primary species	Secondary species
Cavernularia lophyrea	*Cavernularia hultenii*
Evernia esorediosa	*Evernia mesomorpha*
Fulgensia fulgida	*Fulgensia subbracteata*
Lecanora intricata	*Lecanora soralifera*
Letharia columbiana	*Letharia vulpina*
Parmelia abessinica	*Parmelia hababiana*
Parmeliella plumbea	*Parmeliella atlantica*
Pseudevernia intensa	*Pseudevernia consocians*

CHAPTER FOUR

DISPERSAL, ESTABLISHMENT AND GROWTH

4.1 Ascospore discharge and dispersal

Ascospores are discharged forcibly in most lichens, but in some genera of the Caliciales (e.g. *Calicium, Sphaerophorus,* see figure 3.1*G*) they form in a powdery mass. Methods of ascus dehiscence have already been considered in section 3.4. Most lichens have perennial ascomata, and discharge appears to be greater in winter than in summer, wet weather favouring discharge. In laboratory experiments ascospore discharge has been found to occur in the dark. In one study of seven unrelated species, 55–76 (mean 66) % of the spores discharged were released in the dark. Diurnal periodicity has been observed in *Huilia macrocarpa* (figure 4.1). Ascospores are discharged only a few mm, with occasional spore masses reaching more than 3 cm. The discharge distance is related to ascus structure, the species, and spore mass size. A single large ascospore or eight adhering small ones will be projected further than single small ones. As asci

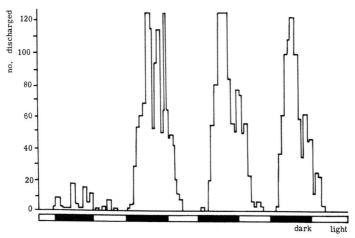

Figure 4.1 Rhythm of ascospore discharge in *Huilia macrocarpa* under alternating 12 h dark/12 h light regimes at $14 \pm 1°C$. Adapted from F.B. Pyatt (1974), in *The Lichens* (V. Ahmadjian and M.E. Hale, eds.), Academic Press, New York and London, 117–145.

45

generally only send spores above the boundary layers into turbulent air, dispersal by other means (wind, water or animals) is then necessary. Stalked ascomata in *Calicium* (figure 3.1) and erect *Cladonia* podetia (figure 2.4) mainly aid dispersal by raising spores into more turbulent air.

4.2   Ascospore germination

Initial ascospore germination and hyphal growth is comparable in speed to that of other fungi. Spore germination does, however, seem to be rather variable with maxima in winter months; it may range between 4% in the summer and 90% in the winter. Ascospores can germinate on a wide variety of artificial media, including nutrient agar, and some may germinate even on clean glass surfaces. Ascospore viability may be reduced after desiccation. In *Huilia macrocarpa* a reduction from 75% to 12% after three weeks' desiccation is reported. In lichens with multinucleate spores, such as *Pertusaria* species, numerous germ tubes may emerge from each spore. A limited amount of hyphal growth arises around the ascospore, but the life of the mycobiont mycelium in the field before lichenization is obscure, although it has been observed in *Dimelaena oreina*.

4.3   Photobiont reproduction and dispersal

In order for the germinating ascospore to develop into a lichen, it must acquire a suitable photobiont. Photobiont sexual reproduction has not been observed in nature, although *Trebouxia* has been observed to form zoospores rarely in the lichenized state (see section 1.6). Photobiont cells may be acquired (1) from free-living algae (although *Pseudotrebouxia* and *Trebouxia*, the commonest photobionts, are very rare in this state), (2) from existing thalli of other lichens (see section 1.4), or (3) vegetative propagules (e.g. soredia) of other lichens.

Propagules including both mycobiont and photobiont (see section 3.7) overcome this problem and place species at an evolutionary advantage (see section 3.8). In a few genera (see section 3.1) photobiont cells live and reproduce asexually *within* perithecia and appear to become attached to discharging ascospores, so being dispersed along with them.

4.4   Photobiont recognition and selection

Mycobionts have a substantial degree of specificity for photobionts (see section 1.6). There are two possible mechanisms by which a mycobiont can recognize an appropriate photobiont.

(1) Recognition of the photobiont cell surface by specific antigens. Phytolectins have been isolated from mycobionts which bind to the cell surface of non-lichenized photobionts. Since the binding is confined to the strains of symbionts specific to the lichen species the phytolectins may enable hyphae to recognize an appropriate photobiont on contact.

(2) Microscopic observation of the establishment of lichen associations in cultures, mainly by Ahmadjian and co-workers over the last 25 years, indicates that recognition is not photobiont-strain specific (see section 1.6). Mycobiont hyphae envelop a variety of candidate photobiont cells, inappropriate ones being killed by *parasitism* (necrotrophism) of the mycobiont while cells tolerating the mycobiont attack can enter the mutualistic system. This indicates that specificity is the result of resistance to parasitism and not recognition by contact.

These alternative explanations have not yet been integrated into a general hypothesis, but the photobiont's specific surface antigens to which the mycobiont lectins bind, could have a role in photobiont survival as opposed to recognition. When mutualistic integration has occurred, ultrastructural changes are soon initiated (see section 5.5.1) and contacts secured.

Ahmadjian's studies have detailed the initial contact events, the developing thallus and the formation of reproductive structures of several lichens in culture under different conditions. The North American *Cladonia cristatella* has proved especially suitable for such experimental syntheses (figure 4.2). In general terms thallus differentiation only occurs when there is an appropriate photobiont. Mycobiont hyphae first envelop a group of photobiont cells, and then lichen tissues begin to differentiate.

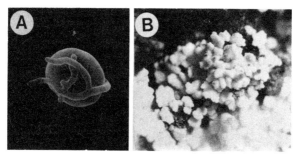

Figure 4.2 Synthesis of *Cladonia cristatella* in pure culture. *A*, Mycobiont hyphae clasping a cell of *Trebouxia erici*, scanning electron micrograph, × 3000. *B*, Synthesized squamules, × 15. *A* from V. Ahmadjian and J.B. Jacobs (1981), *Nature* **289**, 169–172; *B* by V. Ahmadjian.

Some structural and ultrastructural differences between natural and resynthesized thalli have been noted. In the natural environment *C. cristatella* may have extensive mycobiont mycelium in the substratum (a prothallus, see section 2.1); hyphae associate with appropriate photobiont cells and quickly differentiate into squamules which later give rise to podetia.

4.5 Asexual propagule dispersal and establishment

Lichens produce several different types of vegetative propagules that contain both mycobiont and photobiont cells (see section 3.7). This method is more efficient than independent dispersal of the partners, and in species pairs the asexual species often exceeds the sexual in its geographical range (section 3.8).

Soredia are sufficiently small to be wind-dispersed, but the dissemination of larger diaspores or thallus fragments may involve water, gale-force winds, or animals, according to the situation. For example, lichens on bird-perching stones are likely to be dispersed by the birds, and those on trees by strong winds or water run-off. After arriving on and attaching to a suitable substratum, soredial growth is quickly followed by tissue differentiation. Some propagules, such as isidia, are already differentiated before attachment. In some cases several young thallus initials formed from separate propagules may unite to form a single thallus.

4.6 Thallus growth

Studies on the growth of lichens have concentrated on species with circular crustose or foliose thalli, and on erect fruticose species. Growth in circular thalli, being mostly two-dimensional, is usually determined by measuring increase in size, while that in fruticose species is measured by weight.

Photography of individual lobes of *Parmelia* species has shown that growth, and/or elongation, takes place in the terminal 2–5 mm with over half in the last 0.6–1.7 mm. Ultrastructural studies of the growing point of *Peltigera canina* have shown that the mycobiont cells within 60 μm of the lobe end have characteristics of cells in a state of rapid division. Concentric bodies, ellipsoidal structures about 350 μm across with a membrane-free central core and an outer layer of radiating membrane plates, which are common in lichenized fungi but rare in other groups, appeared in great numbers in this zone and were then distributed, probably passively, in the mycobiont during differentiation. The photobiont cells were also apparently in a state of division at the lobe end. Behind the zone of cell division, the mycobiont undergoes differentiation and cell enlargement. It is not clear how the essentially unicellular photobiont is able to colonize and form the photobiont layer of the new thallus in the zone of cell division. (Reports in the older German literature of *Schiebenhyphen* which push *Wanderalgen* forward to the thallus margin have not been subject to critical reinvestigation.) The discontinuous colonization of the thallus margin by the photobiont can be seen as areolae on the prothallus of *Rhizocarpon* and some other areolate species.

As cell division occurs at the growing point and elongation behind it, lateral translocation of carbohydrate and other nutrients probably takes place to maintain the synthesis of cellular components. Photosynthetic production of carbohydrate per unit area may be presumed to be uniform over the lobe or thallus as a whole, although there is some evidence that the rate of photosynthesis is lower in older parts of the thalli. These assumptions have been used by Aplin and Hill to construct a growth equation that relates the rate of radial growth of circular thalli to the size of the thallus and time:

$$dr/dt = asr/(r + 2s) \qquad (1)$$

where r = thallus radius, a = rate constant for growth and s = a distance constant which accounts for the lateral translocation of carbohydrate. The equation, which has been shown to describe lichen growth in practice (figure 4.4), is similar in concept to the one applied by Trinci for the growth of fungus mycelia on agar:

$$dr/dt = as \qquad (2)$$

where a = rate constant for growth and s = the distance granules of wall material can move forward in the terminal syncytium. Unlike fungal mycelia, in which s can be measured anatomically, lichen thalli have no structural manifestation of s, which is more closely related to a diffusion constant. Values for the rate constants of some lichens and common laboratory moulds are compared in table 4.1. Although the values of the rate constants for lichens are about 1000 times lower than for the moulds, it should be noted that (1) lichens only spend a portion of their time in a state of physiological activity, and (2) many other non-lichen-forming fungi, for example *Lecanidion* and *Odontotrema* species with perennial ascomata, may also have very low growth rates. Nevertheless, slow growth of this order implies special characteristics, such as considerable cell longevity,

Table 4.1 Rate constants for growth in lichen-forming fungi in natural habitats compared with some non-lichen-forming fungi in culture.

	Maximum radial increase $(mm\,yr^{-1})$	Rate constant (h^{-1})	Maximum radial increase $(\mu m\,h^{-1})$
Lichens in natural habitats[1]			
Parmelia conspersa (foliose)	5.5	1.08×10^{-4}	0.63
Diploicia canescens (placodioid)	1.75	1.98×10^{-4}	0.20
Rhizocarpon geographicum (areolate crustose)	0.5	0.31×10^{-4}	0.060
Fungi in culture[2]			Hyphal growth rate $(\mu m\,h^{-1})$
Neurospora crassa (anamorph)		4.40×10^{-1}	4400
Aspergillus niger		1.17×10^{-1}	133
Penicillium chrysogenum		1.53×10^{-1}	76

[1]data from D.J. Hill (1981), *Lichenologist* **13**, 265–287.
[2]data from A.J. Trinci (1971), *J. gen. Microbiol*, **67**, 325–344.

and resistance to environmental hazards, for example browsing by invertebrates, disease resistance and tolerance of climatic extremes, of which faster-growing organisms have less need.

Lichens in taxonomically distant groups tend to form lobes of remarkably similar structure, both in foliose and placodioid species (see section 2.1). One explanation of this phenomenon is that lobes are an adaptation to the contraction and expansion of thalline tissues during desiccation and rehydration. In structurally simple crustose species, the thallus cracks into indeterminate areas, *secondary areolae*, between which cracks open on drying and close on wetting. The cracks form in a random manner and break up the physiological integrity of the thallus tissues (figure 2.1*E*). In more complex crustose species, the thallus redevelops in small discontinuous patches (*primary areolae*) which are unaffected by desiccation and rehydration. In placodioid thalli, primary areolae are contiguous and elongate radially during growth and appear lobe-like at the edge (figure 2.1*F*). The lobes of foliose thalli are detached from the substratum but held down, for example, by rhizines (see section 2.4). The presence of dorsiventral lobes permits the unimpeded radial translocation of nutrients required for growth, but does not permit the tangential translocation necessary for the maintenance of circularity as the thallus enlarges. The circularity of lobe-forming species is maintained by constant lobe division whether or not there is space to accommodate the new lobes. This leads to increasing competition among lobes for space, and increasing lobe extinction as thalli grow larger.

The rate of growth of lichen thalli is subject to considerable variation and apparently identical thalli in an apparently uniform habitat may grow at different rates (see figure 4.3). This variability may be due to small-scale concentrations of nutrient availability, such as bird droppings, as well as variations in microclimate. In *Parmelia saxatilis*, slow growth has been found to be associated with the production of narrow lobes, although there is usually no visible sign of differing growth rate. The growth of lichens is also subject to considerable seasonal variation, with the fastest rates occurring in autumn, winter and spring when thalli spend a greater

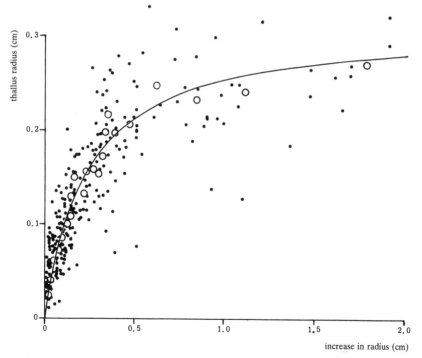

Figure 4.3   The relationship between radial growth (increase in radius over 2 years) and thallus radius in *Diploicia canescens*. The thallus radius is the geometric means of radii at the start and end of the 2-year period. ● = measured thallus; o = blocked mean value. Line is best fitting one using Equation 1. Adapted from D.J. Hill (1981), *Lichenologist* **13**, 265–287.

proportion of the time in a hydrated state. In *Parmelia caperata*, accumulated growth has been found to be directly proportional to accumulated rainfall.

A feature of many crustose and foliose species is that in the older more central parts of the thallus growth continues in one or more ways: (a) increase in thickness, as in *Lecanora atra, Ochrolechia* species, *Pertusaria corallina* and *Peltigera* species; (b) formation of additional lobes or corrugations, as in *Parmelia omphalodes* and *P. pulla*; or (c) production of vegetative propagules and sexual reproductive organs. Thalli may also tend to die out in the centre although marginal growth does not appear to be affected. This is seen in lobate *Caloplaca* species such as *C. cirrochroa*, and also in *Parmelia saxatilis, P. glabratula* and *Xanthoria parietina* in areas affected by air pollution.

The pattern of growth in fruticose lichens may be broadly similar to that in foliose and crustose species. Measurement is usually made in terms of increase in dry weight as it is, for example, possible to remove and weigh tufts of *Cladonia* species without causing significant damage. In *C. stellaris* a direct relationship between relative growth rate and rainfall has been found. Topographically, the thallus of a fruticose lichen is usually tapered from the youngest terminal region to the thicker older parts. Growth can therefore occur equally throughout the thallus and need not be restricted to a clearly marked growing point at the apex. Umbilicate species may be expected to resemble fruticose rather than foliose species in their growth characteristics.

Studies of lichen growth rates have application in lichenometry (see section 9.5).

CHAPTER FIVE

METABOLISM AND PHYSIOLOGY

5.1 Introduction

Lichen thalli have several distinctive physiological characteristics.

(1) Most species are poikilohydric (i.e. able to function or survive at varying water contents) while most other nutritional groups in the fungi are not. This feature, also seen in many bryophytes, enables lichens to occur in habitats where survival from drying out is essential.

(2) Lichens also grow slowly compared with most other organisms, sometimes increasing their biomass only 1% each year.

(3) The mycobiont derives its carbohydrates biotrophically from the autotroph (photobiont), in a way comparable to the situation in mycorrhizal and many obligate pathogenic fungi. Fungi tend to have a high carbohydrate requirement which is met in lichen associations by the absorption products of photosynthesis from photobiont cells in the thallus. Algae also function as biotrophic donors in other mutualistic symbioses, such as dinoflagellates in corals and *Chlorella* in *Hydra*.

(4) Lichens are also tolerant of extreme environments; poorly-lit tropical rainforests, the Antarctic, hot deserts, fresh water and the sea. Comparable abilities to expand into almost all parts of the biosphere are only otherwise seen in bacteria.

5.2 Water relations

Lichen thalli are able to withstand repeated wetting and drying, and their tolerance to drying is much greater than in most other fungi studied. Substantial increase in water content is usually by the absorption of liquid water, although some species have a hydrophobic surface and occur in ecological niches where only water vapour (or liquid water from the substratum) is available (*aerohygrophilous*).

Lichens can absorb sufficient water from damp air to become physiologically active. *Ramalina maciformis*, a species of hot deserts, takes in sufficient water from air with a humidity greater than 80% to photosynthesize; at a relative humidity of 95%, CO_2 uptake is identical to that of a fully hydrated thallus. In general, in a day of humid weather, the CO_2 incorporation of a lichen can be expected to rise to 50–80% of the maximum obtained by adding liquid water directly.

Water loss by drying is much slower than water gain from liquid water taking hours rather than minutes. Drying rates vary enormously and are dependent on air humidity, movement and temperature, radiant energy, thallus morphology, and microtopography.

Water occurs within lichens in three main sites, depending on how much potential energy (water potential ψ) is present compared with pure water and location.

(a) *'Solid state' water* ('held' or 'bound' water). This is tightly held by hydrogen bonding to macromolecules and other components and present even when the thallus is 'air dry'. This water associated with the structure of the cellular components does not permit the macromolecular movement that occurs in 'liquid phase'. Thus no significant physiological activity can take place. 'Solid state' water may be intracellular and extracellular, amounting up to 20% of the oven-dry weight of the thallus. The amount of water depends on the water vapour pressure (humidity) of the air. Its water potential is very large and negative.

(b) *Cell water*. Water in cells contributing to the 'liquid phase' with a high osmotic pressure owing to the cellular substances dissolved in it. The volume of water contained increases until wall pressure potential (ψ_t) exerted by the cell wall equals the osmotic potential (ψ_p) within the cell; no further water can then be taken up by the cells. Water potential for the cells is $\psi_p + \psi_t$ and may have an order of magnitude of to 10^2 bar. This water contributes about 20–80% above oven-dry weight.

(c) *Capillary water*. When $\psi_p = \psi_t$ water may be held within the cell wall and amongst hyphae and the photobiont cells by capillary action. This is also referred to as *matric water* (water potential ψ_m), and can amount to many times the dry weight of the thallus.

In most homoiohydric organisms only the fully hydrated (unplasmolysed) form is normally significant, but poikilohydric organisms such as lichens have contrastingly different structural and physiological states. Some lichen thalli are crisp and brittle when dry and easily break into fragments (see section 3.7). Some crustose lichens may be cracked into areoles when dry (figure 2.1 E); each areole is probably physiologically independent and on wetting swells to form a continuous surface. The transparency of the upper cortical layers of a thallus may also vary with water content, becoming more translucent when wet. In *Phaeophyscia*

Table 5.1 Water contents of some fully hydrated lichens. Data from O.B. Blum (1974) ['1973'], in *The Lichens* (V. Ahmadjian & M.E. Hale, eds.), (Academic Press, New York and London), 381–400.

Lichen genus	Type of photobiont	No. of species studied	Type of thallus	Mean max. water content (% oven dry weight)	Notes
Aspicilia	green alga	2	crustose	139	—
Cladonia	green alga	7	fruticose	153	terricolous
Parmelia	green alga	4	foliose with upper and lower cortex	191	—
Peltigera	cyanobacteria	3	foliose with only upper cortex	298	water held in thick spongy medulla
Collema	cyanobacteria	2	foliose without cortices	606	water held in gelatinous extracellular polysaccharide

orbicularis, for example, the dry thallus is grey and becomes green when wet, the cortex becoming sufficiently translucent for the photobiont cell colour to show through.

A 'dry' lichen has negligible physiological activity, but may survive lengthy periods of desiccation. The mechanism by which the cells survive desiccation to very low water potentials, for example -100 to -1000 bar, or even storage in a P_2O_5 desiccator, while maintaining the structure of macromolecules, is unknown. However, molecules of 'solid state' water associated with macromolecules are mobile and can be replaced to a certain extent by hydroxyl groups on the other molecules, such as polyols. Lichen thalli contain large amounts of polyols, usually amounting to 3–12% of dry weight. The concentration of polyols in the cells will be very high since the cell wall and extracellular polysaccharides comprise a large part of the dry weight of a lichen. Green photobionts also contain polyols to quite high concentrations, although blue-green photobionts do not contain any appreciable soluble carbohydrate. High polyol concentration may have a further role in decreasing water potential, thereby facilitating water absorption from the air and the maintenance of a hydrated state.

Lichen cells, in common with those of most bryophytes but unlike some vascular plants, do not exhibit sudden changes in basic physiological activity over a wide range of water contents. Sudden re-saturation of a dry thallus may, however, result in abnormally high respiration rates (see

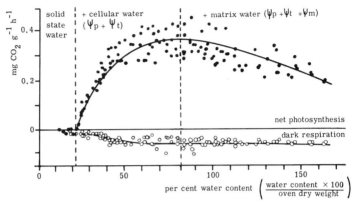

Figure 5.1 Effect of water content on net photosynthesis and dark respiration in *Ramalina maciformis*. Adapted from O.L. Lange (1980), *Oecologia* **45**, 82–87.

below). An outline of the way in which net photosynthesis and respiration respond to thallus water content is presented in figure 5.1. 'Solid state' water (0–22% dry weight) permits negligible activity. 'Liquid phase' cellular water (22–80%) content and physiological activity are directly related and both CO_2 uptake and CO_2 loss reach maximum values. The matric water (80–160%) impedes CO_2 uptake by the thallus in light, as O_2 diffuses down a slight gradient from 0.03% air, but does not impede respiratory loss of CO_2 as the CO_2 is generated throughout the thallus and down a potentially much steeper gradient. A different graph would be obtained if CO_2 exchange against water potential (i.e. $\psi_w = \psi_p + \psi_t + \psi_m$) was plotted. The matric water part would then be foreshortened as there is little energy in the matric water potential; matric water is easily lost, and the other end extended. Water potential would be a useful parameter in data presentation, but is a little more difficult to measure.

Temperature, light intensity and gaseous concentrations naturally limit rates of photosynthesis and respiration, but the effects of low water content are less obvious. In cells with less than their maximum water content, where $\psi_w = \psi_p$, the strongly negative water potential is a drain on the free energy of the system; energy output in terms of physiological rate is therefore reduced. The molecular motion of the system is being reduced because of the lack of associated water molecules, and the energy required to mediate reactions is consequently greater. The effect is similar to that of low temperature.

Prolonged saturation with water appears to be fatal to most non-aquatic lichens, which therefore seem to need a limited supply of water. It is not clear whether periods of dryness are essential, or if constant high water contents are harmful. There is some evidence that low water contents are required for some physiological processes and high ones for others. When a dry lichen is suddenly re-saturated, the following events take place.

(1) A rapid release of gas (apparently CO_2-rich) occurs in the first seconds. This also occurs with glass or filter paper as well as lichens, and is an entirely physical response.

(2) A rise in respiration rate to several times the base rate (measured for a fully hydrated thallus) occurs within a few minutes. It then gradually returns to the base rate over one (*Hypogymnia physodes*) to ten hours (*Peltigera polydactyla*).

(3) A loss of solutes such as polyols and phosphates from the thallus into water in the intracellular spaces. This also occurs within the first 2–5 min in *Peltigera polydactyla* and is due to membrane permeability.

(4) A gradual rise in photosynthesis with time (several minutes or hours in *Peltigera polydactyla*).

(5) The rehydration of the tissues, re-establishment of macromolecular and membrane functions, and consequently the possibility of solute reabsorption.

(6) A net loss of carbohydrate reserves, arabitol in *Hypogymnia physodes* and mannitol in *Peltigera polydactyla*. This may be in the order of 20% of reserves, or in *P. polydactyla* this can be 0.65 mg mannitol cm^{-2}.

If lichens were wetted and dried repeatedly in quick succession without time between the wetting episodes for photosynthesis and mineral absorption to take place, their reserves would gradually be depleted. It takes only a few cycles in the dark to reduce the polyol content drastically. In natural environments such extreme conditions are unlikely as (a) it is unusual for a dry lichen to be suddenly saturated with water and quickly dried out again, especially at night, and (b) in dry periods lichens may have episodes when they are at least partly hydrated, for example by moisture from dew or humid early-morning air, so that a limited amount of photosynthesis can proceed.

The mechanisms by which lichens survive and grow in extremely xeric habitats have attracted the attention of Lange and his co-workers since the late 1960s. They measured the photosynthetic and respiratory gaseous exchanges in *Ramalina maciformis* in the Negev Desert, Israel, at the end of the summer dry period (figure 5.2). In this area the rainfall is negligible

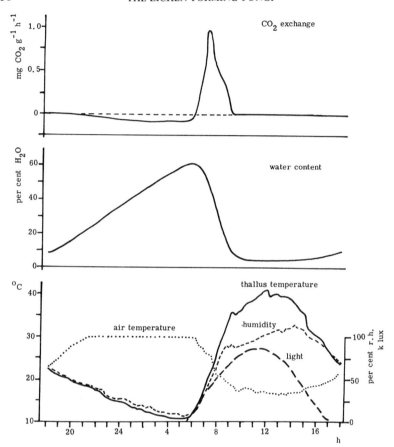

Figure 5.2   Water content and CO_2 exchange in *Ramalina maciformis* in the Negev Desert over 24 h. Adapted from O.L. Lange, I.L. Geiger and E.D. Schulze (1977), *Oecologia* **28**, 247–259.

during summer, but the lichens are able to obtain sufficient water from the dew which frequently forms at night. The thallus hydrates overnight as the air becomes humid and dew forms, and respiration proceeds at a steady rate. At dawn increasing light intensity and temperature cause a rapid rise in photosynthesis. This quickly reaches a maximal value and then declines rapidly owing to the limiting water content; the photosynthesis is therefore first light- and then water-limited. The dry thallus then remains relatively inactive until rehydrated the following night. The overall photosynthetic gain, though variable, amounted to about 1.32 mg CO_2/g dry weight/day, whereas respiratory loss amounted to 0.78 mg CO_2/g dry weight/day. The difference (0.54 mg CO_2/g dry weight/day) could account for the observed annual 5–10% increase in size. In this example, lower winter temperatures would favour a greater uptake of CO_2, as would the increased precipitation (50–150 mm year^{-1}) at that time of year.

5.3 Light and photosynthesis

The mutualistic lichen association integrates the mycobiont and photobiont into a single photosynthetic unit. Although the reactions of photosynthesis are carried out by the photobiont, the mycobiont is directly involved in other associated processes, such as light absorption, CO_2 diffusion and energy balances.

Conditions of illumination experienced by lichens vary enormously according to the habitat. *Clathroporina* species on palm trunks in the understorey of a dense tropical rain forest experience very low light in a warm and damp environment, *Rhizocarpon geographicum* experiences cool high irradiance and dry conditions, in hot deserts lichens spend most of their existence in a desiccated state, and those in the Antarctic in a frozen state. The capacity of lichens to utilize extremes of irradiation and temperature exceeds that of most other plants. The sun's radiation affects a lichen thallus in several ways.

(a) Long-wave radiation (> 700 mm) can have a heating effect but its absorption, irradiance and reflectance is not well understood. An increase in temperature can cause a warming of the air in the boundary layer, thallus temperatures can rise substantially above ambient and may be 50°C in dry thalli. Infra-red reflectance and irradiance therefore appear to convey an advantage to lichens in warm habitats, and infra-red absorbance an advantage to ones in cold habitats. For short periods, comparable temperatures may be experienced by wet thalli, although wet thalli will remain cooler as a result of water evaporation.

(b) Short-wave radiation (< 300 mm) can be very damaging but may be absorbed by pigmentation. Lichen products (see chapter 8) with an aromatic arrangement of carbon atoms that occur in the upper cortex will tend to absorb ultra-violet light, and so shield the underlying photobiont cells.

(c) Visible radiation is frequently substantially reflected and absorbed by the upper cortex. This gives the thallus its colour. Some thalli, noticeably those from poorly illuminated habitats, appear green when wet due to increased visibility of the photobiont (Table 5.2). The cortex

Table 5.2. Thallus parameters in *Xanthoria parietina* from highly illuminated (rocks by sea) and poorly illuminated (tree trunk) habitats. From D.J. Hill and H. Woolhouse (1967), *Lichenologist* **3**, 207–214.

	Dry wt unit area^{-1} (mg cm^{-2})	Chlorophyll (µg cm^{-2})	Physcion µg cm^{-2}	Thallus thickness (µm) (mean)	Upper cortex thickness (µm)	Photobiont layer (µm)
Rocks by sea	10.99 ± 0.45	18.1 ± 1.4	408.1 ± 24.6	198.6	25.4 ± 1.2	34.2 ± 3.2
Tree trunk	7.24 ± 0.68	31.5 ± 2.18	76.6 ± 21.5	137.6	17.6 ± 1.3	46.4 ± 6.0

Table 5.3 Photosynthetic rates of lichens compared with other organisms.

Organism	Photosynthetic rate
	$\mu M\,CO_2\,mg^{-1}\,chl\,h^{-1}$
Cladonia rangiformis	70 − 138 (chlorophyll a)
Peltigera praetextata	414 − 460 (chlorophyll a)
Azolla caroliniana	40 − 100 (chlorophyll)
	$mg\,CO_2\,dm^{-2}\,h^{-1}$
crops	20 − 29
deciduous trees	6 − 20
Hypogymnia	3 − 4

in some lichens can reflect or absorb light of similar frequencies to those absorbed by photosynthetic pigments. *Xanthoria* species appear yellow because substantial absorption at the blue end of the spectrum virtually eclipses blue light from the chlorophyll of the photobiont cells.

(d) Lichen thalli can adapt to different light regimes in a comparable manner to flowering plant leaves. In well-illuminated habitats, thalli tend to have a thicker and (or) more deeply pigmented cortex, and a lower density of photobiont cells and chlorophyll (table 5.2).

Although the mycobiont has an essential role in controlling the light conditions incident on the photobiont cells, the principal photosynthetic mechanisms of the photobionts of lichens are similar to those of free-living algae and cyanobacteria. The rate of CO_2 uptake in lichens is lower per weight of lichen or per unit area than rates found in flowering plant leaves, although on a unit of chlorophyll basis it is quite high (table 5.3).

5.4 Factors affecting photosynthetic rate

Net uptake of CO_2 in the light has generally been used to measure lichen photosynthesis (NAR or net photosynthesis), with dark respiration recorded for comparison. Respiration rates are rapid owing to the large volume of fungal tissue in the thallus, and net photosynthetic rates show pronounced temperature optima as respiration rates reach a limiting value at a higher temperature than do photosynthetic ones. The optimal temperatures are usually in the range 5–20°C, but depend on the species, water content, season, irradiance, and possibly other factors. Under comparable conditions, lichen species from the tropics tend to have higher temperature optima (above 18°C), than ones from the Arctic (below 10°C). Most lichens can still photosynthesize at quite low temperatures (table 5.4).

Table 5.4 Lowest temperatures at which lichens show net photosynthesis. (Data from L. Kappen (1974) ['1973'] in *The Lichens* (V. Ahmadjian and M.E. Hale, eds.), Academic Press, London and New York, 311–380, and D.H.S. Richardson, *ibid.*, 269–288.

Alpine	°C	Tropical	°C
Cladonia foliacea	− 24	*Heterodermia leucomelos*	− 1
Stereocaulon alpinum	− 24	*Parmelia schimperi*	− 3
Cladonia elongata	− 13	*P. africana*	− 12
Arctic/Antarctic		**Temperate**	
Usnea acromelana	− 18.5	*Hypogymnia physodes*	− 6
Cetraria nivalis	− 20	*Lobaria pulmonaria*	− 7
		Usnea ceratina	− 10
		U. submollis	− 10

Temperature optima have been emphasized in ecological studies, but more complete data, also including the effects of water content and irradiance have yielded fuller physiological response patterns.

Photosynthesis shows not only compensation points for CO_2, irradiance and temperature, but also for water content. *Ramalina maciformis* reached a water compensation point when it was in equilibrium with 80% relative humidity ($\psi_w = 287$ bar). The compensation point for any one parameter depends on the magnitude of the others. In view of the large amount of mycobiont tissue, lichens tend to show a very high light intensity compensation point at high temperatures. Lichens on the whole are consequently less well adapted to growth in poorly illuminated situations than mosses. In particularly well-illuminated sites, photosynthesis is probably limited by hydration and the rate of CO_2 diffusion into the thallus. Cyphellae, pseudocyphellae and a pored epicortex have a key role in facilitating this diffusion (see section 2.3). Recent work suggests that diffusive resistances involve the combination of water and the anatomy and morphology of the whole thallus. Indeed, at optimal water contents the resistance may not limit the rate of photosynthesis. A study by Green showed that although CO_2 was thought to be incorporated by ribulose diphosphate carboxylase activity, with a minor role of PEP carboxylase (probably due to dark fixation by the mycobiont), some macrolichens have features of CO_2 fixation (low compensation point and little effect of low oxygen partial pressure) characteristic of C4 rather than C3 photosynthesis.

5.5 Carbon transfer

The mechanism by which the mycobiont obtains photosynthetic products from the photobiont was first elucidated by D.C. Smith and co-workers in

Oxford, beginning in the early 1960s. Extensive studies carried out by this group have stimulated a wider investigation of the transfer of nutrients in other mutualistic symbioses, including algae–invertebrates, mycorrhizas, and obligate fungal pathogens on vascular plants. The methods used in lichens were based on the photosynthetic incorporation of $^{14}CO_2$ followed by analyses to locate and identify labelled components as the carbon transfer proceeded. Carbon transfer from the photobiont to the mycobiont was found to be (see also figure 5.3) (a) rapid, (b) substantial in quantity, and (c) involves the transfer of a single simple carbohydrate, with (d) carbohydrate accumulating in the mycobiont as a polyol.

The transfer of photobiont products does not depend on the mycobiont harvesting dead photobiont cells, but involves live cells. Photosynthetic products can be detected in the mycobiont a few minutes after fixation of $^{14}CO_2$ by the photobiont. In 'pulse-chase' experiments with *Peltigera*, about 70% of the fixed $^{14}CO_2$ appears in the fungus as mannitol. Most of the products of photosynthesis are available to the mycobiont rather than the photobiont. The amount of carbohydrate transferred has been estimated in *Peltigera* to be 4–6 μmoles glucose g^{-1} dry weight h^{-1}. Although small compared with net photosynthetic fixation rates, up to 230 μmoles $CO_2 g^{-1}$ dry weight^{-1}, experimental conditions used in these studies were very unnatural. Thallus samples were floated on aqueous media, causing total saturation with water, and this reduces the photosynthetic rate. However, in *Peltigera polydactyla* transfer can take place at a lower water content when photosynthetic rates are higher, but may be impeded when water content limits the rate of photosynthesis.

The single simple carbohydrate transferred depends on the photobiont. Cyanobacterial photobionts release glucose, while algal photobionts release a polyol. The commonest algal photobiont, *Trebouxia*, releases the polyol ribitol, *Hyalococcus* sorbitol, and *Trentepohlia* erythritol (table 5.5).

In these polyols and glucose, three carbon atoms of each carry identical groups. The significance of this is obscure, but may be related to the specificity of the mycobiont uptake mechanism.

Photosynthetic production by the cyanobacterial photobionts appears to be quite different to that of polyol production by algal photobionts. In *Nostoc* the glucose exists as a small pool in equilibrium with a larger one of glucose as an α, 1–4 glucan. There is an immediate transfer of ^{14}C-labelled glucose molecules, transfer ceasing immediately photosynthesis stops. In algal photobionts there is generally a substantial pool of the polyol in the photobiont cell. Therefore, after fixation, transfer of ^{14}C atoms takes longer than with cyanobacteria; the transferred polyol is not in equilibrium

Table 5.5 Identities of carbohydrates transferred in lichens with different photobionts.

Photobiont	Examples of lichens used in experiments	Carbohydrates transferred between the symbionts
Algae		
Trebouxia	Xanthoria calcicola (syn. X. aureola)	Ribitol
Myrmecia	Lobaria amplissima	Ribitol
Coccomyxa	Peltigera aphthosa	Ribitol
Hyalococcus (= ? Stichococcus)	Dermatocarpon miniatum	Sorbitol
Trentepohlia	Roccella phycopsis	Erythritol
Cyanobacteria		
Nostoc	Peltigera polydactyla	Glucose
Calothrix	Lichina pygmaea	Glucose
Scytonema	Coccocarpia sp.	Glucose

with a polymeric form. These differences result in difficulties when comparing rates and extents of carbon transfer between symbionts of cyanobacterial and algal lichens. Only a trace of the transferred carbohydrate occurs between the symbionts at any one time as the outer membranes of the two symbionts are only separated by a very small distance and the fungal uptake is very efficient.

The carbohydrate is converted rapidly into mannitol and/or polyol, soluble carbohydrate or polysaccharide, in the mycobiont. Polyols seem to act as short-term reserves and ^{14}C label that accumulates in mannitol after fixation of $^{14}CO_2$ by the photobiont gradually disperses in the metabolism of the mycobiont into longer-term reserves (e.g. lipids) as the mannitol pool turns over.

5.5.1   Mycobiont–photobiont contacts

The mycobiont hyphae and photobiont cells must be in intimate contact to permit the transfer of nutrients and photosynthetic products.

Contacts (figure 5.3) may be either *intracellular* and similar to haustoria formed by plant pathogenic fungi in which the photobiont cell wall is penetrated, or *intraparietal* (intramembranous) where varying degrees of invasion occur but in which the photobiont cell wall is not pierced. Intracellular contacts in lichens are mainly encountered in invasions of dead photobiont cells, in structurally simple crustose thalli, or in synthetic lichens produced in culture. The contact types differ considerably in the

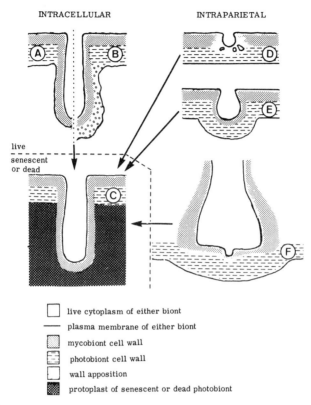

INTRACELLULAR   INTRAPARIETAL

live
senescent
or dead

☐ live cytoplasm of either biont
— plasma membrane of either biont
▨ mycobiont cell wall
▤ photobiont cell wall
☐ wall apposition
■ protoplast of senescent or dead photobiont

Figure 5.3 *A-F*, Types of mycobiont-photobiont contacts (haustoria) in lichens. Adapted from R. Honegger (1984), *Lichenologist* **16**, 111–127.

extent of penetration and this can vary from one thallus (or even part of a thallus) to another within a single species depending on the severity of the environment (figure 5.4).

Ultrastructural studies have revealed remarkable similarities in the molecular arrangements at contact sites. In *Trebouxia* and allied photobionts almost identical tesselated patterns are developed on both mycobiont and photobiont walls from the earliest stages of contact. This almost results in a confluence of the two layers.

Contacts start to be made even while a young photobiont cell is enclosed within its mother-cell wall; the hyphae branch and clasp the photobiont cells (figure 4.2).

The methods of contact provide a greater surface area for the transfer of

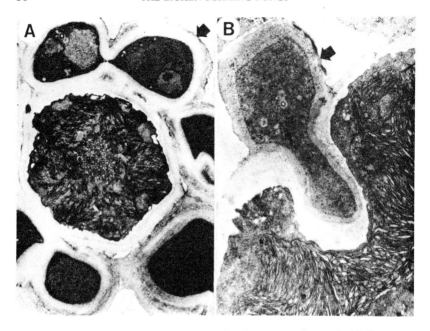

Figure 5.4 Mycobiont-photobiont contacts in *Lecanora radiosa. A*, Mediterranean environment (intraparietal), × 600. *B*, Sinai desert (intracellular), × 11 000. Transmission electron micrographs; mycobiont hyphae arrowed. Adapted from M. Galun, N. Paran and Y. Ben-Shaul (1970), *J. Microscopie* **9**, 801–806.

carbohydrates and other compounds from the photobiont cells to the mycobiont. When aged or dead photobiont cells are invaded by haustoria, the mycobiont is presumably also utilizing products of their natural degeneration. The situation can be interpreted as one of controlled parasitism.

5.5.2 Demonstration of carbohydrate transfer

Carbohydrate transfer in lichens can be demonstrated in the following experiment, based on the fact that lichens exposed to $^{14}CO_2$ in the light transfer the ^{14}C-labelled carbohydrate from the photobiont to the mycobiont. The mycobiont uptake of the ^{14}C-labelled carbohydrate can be intercepted by flooding the lichen thallus with a non-labelled version of the carbohydrate or a competitive inhibitor of mycobiont uptake (figure 5.5). In either case ^{14}C-carbohydrate that would have gone into the mycobiont passes into the medium although the amount of ^{14}C appearing in the

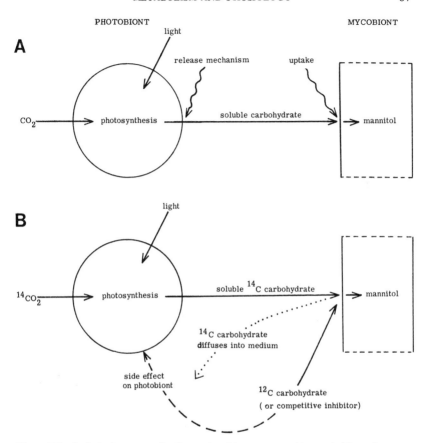

Figure 5.5 Carbohydrate transfer from photobiont to mycobiont. *A*, Normal process (control). *B*, Process in experiment demonstrating transfer ('inhibition technique').

medium as ^{14}C-carbohydrate underestimates the amount of ^{14}C-carbohydrate that would have been transferred to the fungus. Discs are punched from a *Peltigera* thallus with a 5–10 mm cork borer. This provides samples with a standard known surface area. Batches of 5–15 discs are then floated on 5–10 ml distilled water in 25 ml Erlenmeyer flasks. Then 5 μCi of NaH^{14}CO$_3$ is added, the flasks sealed, and placed under light (100 μEm^{-2} sec^{-1}) at 15–20°C for 10 min. Discs are then removed and placed on either 10 ml 0.02 M glucose (or 2-desoxyglucose) or distilled water as a control in a similar flask. Every 20 min an aliquot (e.g. 0.1 ml) of glucose solution and control is removed and the amount of ^{14}C measured either using a planchet counter after drying the sample down, or with a scintillation counter. After about 2 h the lichen is removed from the medium and extracted with boiling 80% ethanol, acetic acid, or 10% trichloracetic acid. The ^{14}C in the extract, and the total amount of ^{14}C leaking into the glucose medium, is then estimated. If using a lichen containing *Trebouxia*, weighed clean thallus tips instead of discs and ribitol instead of glucose are used, and the time in the ribitol solution is extended from 2 hours to 24.

In lichens where it is not known which carbohydrate moves between the symbionts, this approach has been used to identify the carbohydrate. This is based on the assumption that mycobiont uptake is susceptible only to the ^{12}C form of the carbohydrates supplied (table 5.6), that causing the most ^{14}C to appear in the medium probably being the one transferred. This can be confirmed chromatographically by identifying the ^{14}C-carbohydrate released into the medium. Only 8–10 of the 37 or so genera of photosynthetic organisms reported as lichen photobionts (see section 1.6) have so far been investigated (table 5.5 and figure 5.5). As the identity of the carbohydrate is related to the generic identity of the photobiont, the list of carbohydrates transferred may well be extended by further research.

This inhibition technique (table 5.6) has been used to investigate other aspects of the movement of ^{14}C-labelled carbohydrate between the symbionts, but its value is limited as the carbohydrate supplied in the medium interferes with that released by the photobiont. Estimates of carbohydrate release obtained are therefore serious underestimates; the

Table 5.6 Effect of carbohydrates during $^{14}CO_2$ fixation in *Dermatocarpon miniatum*. From D.H.S. Richardson, D.J. Hill and D.C. Smith (1968), *New Phytol.* **67**, 469–486.

Carbohydrate supplied in solution	% of total ^{14}C which appeared in medium
Sorbitol	56.6*
Arabitol	27.2*
Mannitol	7.3*
Glucose	5.4
Ribitol	1.9
Erythritol	1.3
Control (water)	1.1

* ^{14}C shown to be in form of sorbitol by chromatography.

interference increases with time, probably as a result of photobiont uptake of the external carbohydrate.

5.5.3   Uptake and release mechanisms

The mechanism of mycobiont uptake in lichens is probably similar to carbohydrate uptake in non-lichenized fungi. Electron microscope studies have shown that mycobiont hyphae in the photobiont zone have a convoluted plasmalemma and are about $0.5–1.0\,\mu m$ from the algal plasmalemma (see figures 5.3, 5.4). This feature of the mycobiont plasmalemma appears to be similar to that of transfer cells in higher plants. It is therefore probable that the increased surface areas of the plasmalemma aid intensive carbohydrate uptake. In other fungi, carbohydrate uptake across the membranes appears to be associated with the movement of ions (H^+ or K^+). Once in the cell, glucose is immediately phosphorylated by ATP and hexokinase. Polyol uptake presumably follows a similar course.

Carbohydrate release by the photobiont is less well understood. In experiments where the thallus is homogenized and the photobiont cells are separated by centrifugation, the photobiont soon loses its ability to release carbohydrate. The 'cleaner' the preparation, the more rapidly it stops, carbon released no longer being in the form of a mobile carbohydrate. Photosynthate is diverted into 'insoluble' material suggesting that the fixed carbon is being used either to build up reserves in the photobiont, or for photobiont growth, or both (see table 5.7). This suggests that the mycobiont exerts some specific control on the photobiont cells which leads to them releasing carbohydrate. However, the mechanism of that control and how the release is mediated is unclear; although hypotheses have been proposed, few are based on experimental evidence.

5.6   Nitrogen fixation

Nitrogen fixation in lichens is restricted to those containing cyanobacterial photobionts, either as the sole photobiont or in more complex three- or four-membered symbioses (see section 2.6). All cyanobacteria-containing lichens studied fix nitrogen, and there are no reliable reports of other lichens showing this ability.

Nitrogen fixation is a particularly energy-demanding process. Consequently, nitrogen can only be obtained from the atmosphere by utilizing an abundant energy source, such as solar energy. Photosynthesis in cyanobacterial photobionts has to supply not only energy by phosphory-

Table 5.7 Effect of photobiont isolation on the fate of photosynthetically fixed carbon. (Data from D.C. Smith (1974), *Symp. Soc. exp. Biol.* **28**, 437–508).

Fate of ^{14}C (as % fixed)	Photobiont condition	Nostoc (Peltigera canina)	Hyalococcus (Dermatocarpon miniatum)	Coccomyxa (Peltigera aphthosa)	Trebouxia (Xanthoria calcicola, syn. X. aureola)
Released from photobiont into medium or trans- ferred to fungus	in lichen	60	55	65	40
	isolated for 0 h	15.3	26.1	23.1	8.0
	isolated for 24 h	7.0	6.2	3.6	1.0
	cultured	4.0	1.3	1.4	2.5
Incorporated into 80% ethanol insoluble fraction (macromolecule)	in lichen	9.0	2.0	21.0	1.0
	isolated for 0 h	27.8	29.6	56.2	35.3
	isolated for 24 h	50.7	40.2	62.8	53.2
	cultured	46.0	50.3	50.8–72.3	59.7

lation but also considerable quantities of fixed carbon as a reductant to reduce dinitrogen. Nitrogen fixation must also be commensurate with the ability of the photobiont to supply the large quantities of carbohydrate required for polyol accumulation and growth of the mycobiont as well as for the photobiont. Lichens also appear to require large polyol pools to withstand desiccation and re-wetting cycles.

The rates of N_2 fixation in lichens with cyanobacteria as the sole photobiont are substantial and have been estimated in *Peltigera rufescens* (photobiont *Nostoc*) to be within the range 4–38 μg N μg^{-1} N day^{-1} and in *Lichina confinis* (*Calothrix*) 0.4–29 μg N μg^{-1} N day^{-1}. There is a wide variation in the rates observed, owing to many factors, including uncritical use of the acetylene reduction technique which has otherwise greatly assisted the study of N_2 fixation in lichens in recent years. The most marked factors are not only water content, irradiance, pH and temperature, but also time of day and time of year. The effect of water content, irradiance and temperature is slightly different for N_2 fixation, from that for CO_2 exchange. N_2 fixation starts slightly later when a dry thallus is re-wetted (about 1 h) and ceases before net CO_2 uptake in a drying lichen; at saturating water content it does not drop in the same manner as CO_2 uptake. The effect of irradiance is complicated by N_2 fixation occurring at a lower but significant rate in the dark. It may be 50% of the rate in light, but is variable depending on the carbohydrate reserves available. Temperature

Table 5.8 Frequency of heterocysts and nitrogenase activity of cyanobacteria in lichen thalli. From data in C.J.B. Hitch and W.D.P. Stewart (1973), *New Phytol.* **72**, 509, and C.J.B. Hitch and J.W. Millbank (1975), *New Phytol.* **75**, 239.

Lichen species	Heterocyst frequency (% of cells)	N_2 fixation (nM/C_2H_2 mgN^{-1}h^{-1})	N- content (mean total N as % dry weight)
Lobaria laetevirens (cephalodia)	30.4	615	—
Peltigera aphthosa (cephalodia)	21.1	1020	2.7–6.4
Placopsis gelida (cephalodia)	15.0	200	1.7–2.6
Collema auriculatum (thallus)	2.7	15	3.6
Lobaria scrobiculata (thallus)	3.9	11	2.8
Peltigera polydactyla (thallus)	5.8	102	3.5
Placynthium nigrum (thallus)	2.0	7	2.7

optima are also somewhat higher (25–30°C) than for photosynthesis in *Peltigera*. Seasonal variation is large with much lower rates in summer and higher ones in winter ($1-3 \, nM \, C_2H_2^{-1} \, mg$ dry weight h^{-1} in summer, and $10-15 \, nM \, mg^{-1}$ dry weight^{-1} in winter for *Peltigera*).

The rate of N_2 fixation by the *Nostoc* photobiont cells in *Peltigera* is 2–3 times greater than that in free-living *Nostoc* cells. In lichens with a single cyanobacterial photobiont, the frequency of heterocysts, specialized cells which fix N_2, is 3–4% of all the cyanobacterial cells. This is about the number to be expected in a culture of free-living filamentous heterocystous cyanobacteria (but see also cephalodia below).

Nitrogen fixation also occurs in discrete cephalodia (see section 2.6). In comparison with thalli containing cyanobacteria as the main photobiont, cephalodia fix nitrogen at a rate disproportionately faster than their size (table 5.8). Associated with this high rate of fixation is a heterocyst frequency much higher than that in free-living cyanobacteria or where cyanobacteria are sole photobionts. In other mutualistic associations of cyanobacteria with autotrophs, regarding a lichen thallus with a green photobiont as an autotroph, *Azolla*, certain liverworts and *Gunnera*, similarly high heterocyst frequencies occur. In these systems, and in cephalodia, it was originally thought that the high rate of N_2 fixation could probably not be supported by photosynthesis in the relatively fewer vegetative cells of the cyanobacteria alone and that the 'autotroph' might well supply reduced carbon compounds to the cyanobacteria. There is little direct evidence of this in lichens although the phenomenon does occur in

Table 5.9 NH_4^+ assimilating enzyme activities in *Peltigera* species. From W.D.P. Stewart, P. Rowell and A.N. Rai (1982), in *Nitrogen Fixation* (W.D.P. Stewart and J.R. Gallon, eds), Academic Press, London and New York, 239–277.

	Enzyme activities ($nM \, mg^{-1} \, protein \, min^{-1}$)				
Cyanobacteria	GS	GDH	ADH	GOGAT	GPT
Peltigera canina					
photobiont (*Nostoc*)	4.1	2.4	5.5	16.8	—
free-living *Nostoc*	72.4	0.0	22.9	40.4	—
Peltigera aphthosa					
cephalodia (contains *Nostoc*)	2.0	398.0	8.5	20.0	40.1
free-living *Nostoc*	60.0	2.0	11.0	25.0	4.8

GS	= glutamine synthetase
GDH	= glutamate dehydrogenase
ADH	= alanine dehydrogenase
GOGAT	= glutamine-oxoglutarate aminotransferase
GPT	= glutamate-pyruvate aminotransferase

other nitrogen-fixing symbioses with autotrophs (see below). The vegetative cells of the cyanobacteria have a very low level of glutamine synthetase (GS) (table 5.9) so that the normal route by which NH_4^+ produced by the nitrogenase complex is incorporated is not fully operational. As no other alternative method of NH_4^+ incorporation exists, NH_4^+ ions diffuse from the cyanobacterial cell and are absorbed by the mycobiont. Most fungi prefer NH_4^+ as a source of nitrogen, the NH_4^+ being assimilated into glutamate by glutamic dehydrogenase. Amino acids arising from the transamination of glutamate are then available for transfer elsewhere in the thallus. In ^{15}N tracer studies using *Peltigera aphthosa*, the amount of ^{15}N in the cephalodia has been found to reach a maximum after about 10 days, with only a fraction of the total ^{15}N fixed remaining in the cephalodium, the rest being exported to the thallus (figure 5.6). Some

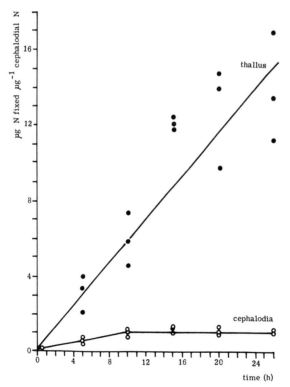

Figure 5.6 Incorporation of ^{15}N into external cephalodia and transfer to the thallus in *Peltigera aphthosa*. Adapted from J.W. Millbank and K.A. Kershaw (1969), *New Phytol.* **68**, 721–729.

appears in the cells of the green photobiont (*Coccomyxa*) but it is interesting that this amount is less than might be expected.

Nitrogen supply from the cyanobacterial photobiont in cephalodia seems to depend on the mycobiont controlling the synthesis of the GS protein in the photobiont. How this is regulated is unclear but is of considerable interest in understanding the lichen symbiosis.

Although nitrogen fixation in the cephalodia of *Peltigera aphthosa* is light-dependent, its extent during dark periods is greater than in free-living *Nostoc*. There is evidence that, since the cyanobacterial photobionts do not grow in the cephalodium once it has fully developed and the N_2 fixed is exported rather than used by the cyanobacteria, CO_2 fixation leads to carbohydrate accumulation as polyglucose (alpha granules) in the light. This store of energy is then available for N_2 fixation in the dark. Consequently, although there is an exceptionally high heterocyst frequency in the cephalodia, N_2 fixation can be supported by the vegetative cyanobacterial cells without the need for carbohydrates from elsewhere in the thallus. Indeed, cephalodia can export some carbohydrate (glucose) to the mycobiont.

In conditions of limiting water supply, N_2 fixation in cephalodia declines (figure 5.7) in a parallel way to photosynthesis. In *Stereocaulon paschale* at saturating water contents, when the rate of CO_2 fixation was sub-optimal,

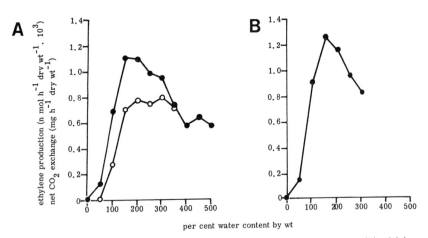

Figure 5.7 Effect of different levels of thallus water content on nitrogenase activity (o) in *Stereocaulon paschale*. *A*, Thalli exposed to acetylene. *B*, Thalli not exposed to acetylene. ● = net CO_2 exchange. Adapted from P.D. Crittenden and K.A. Kershaw (1983), *New Phytol.* **80**, 393–401.

N_2 fixation maintained a maximal rate. As water is lost, N_2 fixation declined more rapidly than and ceased before CO_2 fixation. Pre-treatment with desiccants, or frost, has a deleterious effect on N_2 fixation.

5.7 Mineral nutrition

Lichens do not normally accumulate excessive amounts of minerals, and their low growth rate (see section 4.6) is associated with a low rate of net accumulation of mineral nutrients. As the experimental application of mineral nutrients does not result in massive increases in growth rates it is probable that lichens usually have a relatively limited nutrient demand. In *Hypogymnia physodes*, for example, the amount of phosphate in rainwater has been calculated to be sufficient for its need in growth. Rainwater and run-off are expected to be sufficient for most lichens, although the substratum may also contribute minerals in some species (see sections 6.1, 9.2).

In addition, rain will leach nutrients from thalli, particularly potassium and nitrogen. Others, including phosphate, may be lost during the rewetting of dry thalli in the period before cell membranes are fully functional (see section 5.2). The gross minimal need of a lichen may therefore considerably exceed the net gain.

The concentrations of nutrients in lichens vary considerably (table 5.10), rendering generalizations difficult. Minerals are accumulated by the following methods.

(1) Uptake by cells (figure 5.8A). The minerals passing across the outer

Table 5.10 Ranges of mineral contents of lichens ($\mu g\,g^{-1}$ dry weight) in unpolluted areas. From E. Nieboer, D.H.S. Richardson and F.D. Tomassini (1978), *Bryologist* **81**, 226–246.

Some essential* cationic elements		Some essential* anionic elements		Some non-essential* elements	
K	500–5000	N	6000–50 000	Al	300–400
Mg	100–1000	P	200–2000	Ti	6–150
Ca	200–40 000	S	50–2000	V	0–10
Fe	50–1600			Cr	0–10
Na	50–1000			Sr	0–700
Mn	10–130			Cd	1–30
Cu	1–50			Hg	0–1
Zn	20–500			Pb	20–100
Mo	0–3				
Ni	0–5				

* For vascular plants (not known for lichens).

membrane to the mycobiont and/or the photobiont. In the cells they either remain in solution or are immobilized as components of some macromolecule.

(2) Binding cations to cell walls (figure 5.8*B*) and other extracellular excretions; some bryophytes, such as *Sphagnum*, also have this ability.

(3) Trapping particulate material containing minerals in the thalli. When mineral analyses are carried out on whole thalli, it is usually impossible to remove all debris, and adhering particulate material, which can account for a substantial proportion of the total content, may be trapped in the thallus (see figure 6.2). Particulate accumulation, for example rock crystals and plant debris, may be an important means of acquiring minerals in some lichen species if they subsequently become solubilized since only nutrients taken up by the cell are of direct importance for growth. Minerals bound onto or contained in extracellular material are also only useful to the lichen if they are subsequently taken up intracellularly. Extracellular polysaccharides and other compounds function rather like ion exchange resins in their ability to retain cations. Indeed, when treated with dilute acid (0.1–0.01 N) they are converted into H^+ form. When lichens are placed in a suitable solution of a salt, for example copper sulphate, nickel chloride or potassium nitrate, the cation will be adsorbed. The adsorbed cation can then be eluted by further treatment with acid or displaced by an element along the electrochemical series (or one of close electrochemical potential but at a higher concentration) (figure 5.8).

Adsorption onto the cells is difficult to measure, but there is evidence in

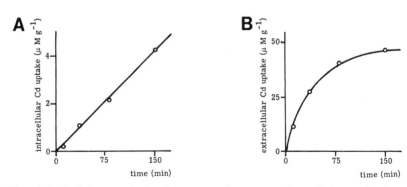

Figure 5.8 Cadmium uptake by *Peltigera membranacea*. *A*, Intracellular uptake (non-exchangeable). *B*, Extracellular absorption (exchangeable). 100M $CdSO_4$ and Cd displaced with 20mM $NiCl_2$. Adapted from R.P. Beckett (1984), *J. exp. Bot.*, in press.

contrast to some metals that 90% of the total potassium absorbed may be intracellular. Intracellular uptake is linear and slower than extracellular adsorption, which is rapid and soon becomes saturated. While extracellular uptake is a physicochemical process, intracellular uptake appears to be metabolic. For example, metal uptake has been shown to be light-stimulated in *Peltigera*. Whether cations adsorbed extracellularly gradually enter the cells over a longer period of time requires confirmation.

Lichens have the ability to absorb minerals intracellularly from dilute solutions and accumulate them with some efficiency. *Hypogymnia physodes* can absorb phosphate as rapidly as yeasts and barley roots. The mycobiont is probably the partner that takes up most of a lichen's minerals. However, the mechanism by which the photobiont obtains sufficient minerals is often discussed in terms of whether the supply of mineral nutrients, on a cellular basis, can be a means by which the mycobiont limits photobiont growth. The photobiont could probably acquire sufficient nutrients from run-off water or those leached out of the mycobiont hyphae, for example following the re-wetting of a dry thallus. The extent to which mineral nutrient loss resulting from re-wetting contributes to the rate of turnover of minerals in the thallus is unclear.

Absorption of minerals directly from the substratum, in addition to those in rainwater and run-off, may also occur. Some organic acid excretions can solubilize cations from rock minerals. This is considered further in relation to the weathering of rocks and minerals in section 6.4. In some ecological niches, lichen thalli may also act as centres of water evaporation, drawing up solutes through the thallus to the surface where they are deposited.

5.8 Heavy metals

Lichens, and also some non-lichen-forming fungi, are able to accumulate metals to high concentrations (0.1–5% of their oven-dry weight). Particularly high contents can be found where the lichen is growing on naturally occurring metalliferous substrata (table 5.10) and especially through accumulation as a result of air pollution (see section 9.2). The amounts of metal found in lichens vary a great deal. When the lichen contains a large amount, it may be in a particulate form or bound to extracellular material or intracellularly located. How much is present within the cells of the mycobiont or the photobiont is not easy to ascertain. Only the intracellular component has any direct effect on the physiology of the lichen.

There is considerable variation in the physiological toxicity of 'heavy' metals. Nieboer and Richardson have conducted extensive investigations in this field since the 1970s. They have classified metal ions into three groups according to their physico-chemical properties. *Class A* includes the alkaline earth elements (e.g. K^+, Sr^+) which are relatively non-toxic and tend to form ligands with hydroxyl and other oxygen-containing sites on organic molecules. *Class B* includes some of the well-known 'heavy metals' (e.g. Ag^+, Hg^{2+}, Cu^+) tending to form ligands with nitrogen- and sulphur-containing groups. A *'borderline'* group includes Zn^{2+}, Ni^{2+}, Cu^{2+}, and Pb^{2+} which form ligands with either. The toxicity of metal ions may therefore be caused by (i) interference with essential groups, such as the sulphydryl groups on proteins, (ii) displacement of normally essential metal ions, and (iii) modifications of the conformational structure of macromolecules.

The principal physiological effects of class B metal ions, and borderline ones with class B properties, are found to be (a) K^+ loss from the thallus, which may be due to interference with membrane functions, (b) a reduced photosynthetic rate (as indicated by CO_2 fixation—photosynthetic O_2 evolution is very sensitive to 'heavy metal' ions), and (c) sensitivity of respiration (usually at slightly higher ion concentration). Photosynthesis in lichens containing cyanobacterial photobionts is much more sensitive to zinc ions than those containing eukaryotic photobionts. Strains of *Peltigera* sampled from metal-contaminated field situations display increased resistance to added Zn and Cd. At least some of this resistance may be phenotypically acquired because it is possible to increase the tolerance of lichens by pre-treating them with low concentrations of metals for several days.

In class A ions, and borderline ions showing class A characteristics, effects on metabolism occur at much higher concentrations than with class B ions.

The relative toxicity of metals on CO_2 fixation in *Umbilicaria muhlenbergii* and *Stereocaulon paschale* has been found to be Ag, Hg > Co > Cu, Cd > Pb, Ni (short-term responses) and Ag, Hg > Cu ≥ Pb, Co > Ni (longer-term responses). Metal ions also cause changes in the response to SO_2 so that for example Cu^{2+} enhances the effect of SO_2, causing K^+ loss, and Sr is antagonistic to SO_2.

5.9 Ecophysiology

The measurement of physiological parameters in lichens in the field is usually very difficult. As most lichens cannot be grown in controlled environments, studies on whole-thallus physiology largely rely on freshly collected specimens. Lichens are very susceptible to small-scale micro-climate, indeed, thallus morphology can create local microclimates, and this can be critically important in lichen-dominated ecosystems. In cold environments, where vascular plants are limited by soil temperatures, lichens are more responsive to localized conditions in the thallus. Surface temperatures, the thallus and first few mm of air above it (the 'boundary layer'), can rise steeply during insolation while air temperature remains low. This may cause a substantial water deficit in vascular plants; frozen roots lead to stomatal closure and hence a temperature increase may not lead to increased photosynthetic activity, but in poikilohydric organisms (see section 5.1), rates of photosynthesis and other cellular functions are raised so long as the water potential does not drop too far. In the tundra, a surface temperature recorded as reaching towards 10° above air tempera-ture could well give rise to a doubling of the rate of carbon gain by the lichen. Similarly, in dry habitats, not exclusively deserts but also in xeric situations on rocks and roofs in temperate latitudes, microclimate can affect not only the amount of dew formed but also the rate at which thalli dry out. In such habitats the total time spent in the hydrated state can be of critical importance in limiting species distribution, for example on north- and south-facing aspects.

In fruticose lichens, which stand out from or hang down away from the boundary layer of the substratum, temperature and humidity of the immediate environment of the thallus will be much closer to meteorological observations.

In the 1950–1960s, Ried pioneered the study of the physiological adaptation of lichens to their habitat by studying the zonations of species in and around streams in the Alps. The aquatic *Verrucaria elaeomelaena* showed its maximum net photosynthetic rate only in fully saturated thalli; after drying for 24 h and re-wetting, respiration rose to six times its normal rate, and the lichen was not able to recover fully. Furthest away from the stream, *Umbilicaria cylindrica* showed its maximal rate of net photo-synthesis at 65% water saturation, and on re-wetting dry thalli of *Rhizocarpon geographicum* only had double the normal respiration lasting for about 1 h. *Huilia soredizoides*, intermediate in distribution between the preceding groups, showed intermediate characters. Further studies led Ried to stress the importance of respiration in the ability of lichens to recover from desiccation and suggest that this ability was one of the most important parameters determining lichen distribution at this site.

Studies in North America, particularly by Kershaw and his co-workers, have used gaseous exchange to measure the response of photosynthesis (and nitrogen fixation) to light, moisture and temperature in the laboratory. The data emanating are presented as a 'gas exchange matrix' (figure 5.9). In this example the overall effect of five parameters, 2 types of lichen material, 2 light intensities, 3 temperatures, 4 samples for seasonal variation, and numerous water contents, on gas exchange can be seen together and their relationships observed. Such detailed data enable predictions to be made as to how the thallus might respond to small-scale microclimatic variation and perhaps to deduce what the characteristics of photosynthesis are, in this example, in shade-adapted thalli compared with ones from open sites. Water content maxima are more or less constant, except possibly with seasonal variation at 15°C, although the effect of temperature is not marked in the open-site specimens unless accompanied by high irradiance, whereas in the shade material it is. At 5°C, photosynthesis appears to be limited by temperature and additional irradiance is

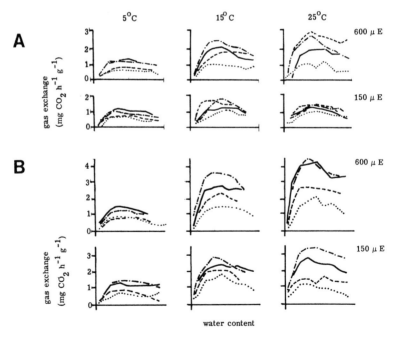

Figure 5.9   Gas exchange matrix for material of *Cladonia stellaris* from different habitats. *A*, Sun ecotype. *B*, Shade ecotype. Graphs within each block are for different times of the year. Adapted from J.D. MacFarlane, K.A. Kershaw and N.R. Webber (1983), *New Phytol.* **94**, 217–233.

of no significance unless accompanied by an increased temperature. Chlorophyll analyses confirmed that the shade specimens contained more chlorophyll, and that CO_2 uptake per unit chlorophyll was the same in shade and open-site material over the range of water contents. Nevertheless, the shade specimens seem well adapted to low irradiance with a CO_2 uptake at 150 μE similar to that of open-site ones at 600 μE.

Similar adaptation has also been documented for *Parmelia sulcata* in trees in Devon. Thalli on the lower part of the trunk had more photobiont (*Trebouxia*) cells per unit area and a higher CO_2 fixation rate compared with higher thalli. Although the adaptation to different environments has primarily been studied by CO_2 gas exchange, and so mostly relates to photobiont behaviour; mycobiont adaptation presumably also occurs. Indeed, the mycobiont may have a role in the organization of the photobiont into different cell densities or thallus shapes.

Larson has found that different *Umbilicaria* species have individual gas exchange matrix patterns that show adaptation to the habitats in which they occur.

The nutrient requirements of different species are likely to vary over a wide range. As in other groups of plants, some are only found in nutrient-rich sites such as bird-perching stones; these may be expected to have a higher demand for, and tolerance of, nutrients than ones from oligotrophic habitats. The limitation of growth by mineral nutrients has been demonstrated by the elevated rates resulting from increased nutrient inputs, for example weekly application of a 2% solution of bird excrement causing a 10–30% increase in growth and twice-monthly application of minerals causing a 40% increase.

Lichens occurring in maritime situations may have different mineral nutrient requirements or tolerance. The mineral nutrition of calcicole and calcifuge species may be contrasting, but it is not known whether aluminium toxicity on acid rocks, and the calcium inhibition of iron uptake on calcareous rocks, is the means of exclusion of the different species concerned as is the case with vascular plants.

5.10 Seasonal variation

Seasonal variation in lichens has been found in various parameters, including (1) dry weight per unit area, (2) growth rates, (3) photosynthetic CO_2 uptake rates, (4) glucose absorption and (5) ascospore dispersal. Two distinct factors are involved, changes directly attributable to environmental factors, and responses to environmental variables indirectly attributable to physiological and structural changes. Changes in physiological behaviour are a type of acclimatization, or acclimation to changed parameters. This was first demonstrated in lichens by M.G. Stålfelt in 1939, but has recently been more extensively studied by Kershaw and his co-workers who have observed acclimation in both CO_2 gas exchange and in nitrogen fixation. The phenomenon appears to be widespread in lichen species, but is not universal.

5.11   Tolerance of temperature extremes

Many lichens can tolerate temperatures below those naturally occurring in the biosphere. Considerable frost resistance occurs in species of cold climates, and even in some with tropical distributions. One arctic-alpine species, *Alectoria ochroleuca*, is reported to have had its basic photosynthetic and respiratory responses unchanged after $3\frac{1}{2}$ years storage at $-60°C$. Most lichens are similarly tolerant of high temperatures, but it is only in the dry state that they can tolerate temperatures above those which occur naturally. When moist, high respiration rates and general disintegration occur at high temperatures. When dry thalli at high temperatures are moistened, respiration is rapid, but the thalli then usually either dry out again before too much damage is done, or cool down as a result of the effect of rain and loss of latent heat of evaporation. Ecological niches in which lichens experience high temperatures are also associated with high insolation, where ultra-violet radiation can be damaging. Lichens in such habitats usually have substantial deposits of pigments, or uncoloured substances, in the upper cortical layers of the thallus which absorb visible and ultra-violet light (see section 8.6).

As in the case of desiccation tolerance, it is not known how lichens are able to tolerate extremes of temperature. However, this does enable lichens to dominate both dry and cold deserts, something also facilitated by their ability to maximize the physiological advantages small-scale microclimates can offer.

CHAPTER SIX

ECOLOGY AND SOCIOLOGY

The critical establishment phase of lichen thalli was reviewed in Chapter 4, and physiological responses to different environmental aspects in Chapter 5. Additional aspects of the interaction of individuals with the substratum and environment (*ecology; autecology*) and of the development and interaction of communities (*sociology; synecology*) are considered here.

6.1 Limiting substratum factors

Several features of the substratum limit the ability of a particular species to develop on it. *Texture* is of major importance in the early stages of the colonization of twigs and smooth pebbles, but some species are especially adapted to such habitats, such as *Arthopyrenia fallax* and *Lecidea erratica* respectively. Conversely, on particularly coarse and hard rocks such as granites with large feldspar crystals, macrolichens such as *Pseudephebe pubescens* and *Umbilicaria cylindrica* are particularly able at overcoming such an uneven substratum. Friable soils are inhospitable to most lichens but colonized by species such as *Toninia caeruleonigricans* and *Psora decipiens* which have developed specialized methods of binding loose soil particles together (figure 2.8).

The persistence of the substratum is also a key factor. Short-lived habitats are less easily colonized; for example, almost no species occur on the leaves of annually deciduous trees in temperate zones while those of evergreen tropical trees can support extensive floras—over 300 species are now known from this habitat.

Water content or retentive capacity may have effects on communities developed, especially on bark, but there is little critical work on this aspect. Even small differences in hydrogen-ion concentrations (pH) are often correlated with conspicuous floristic differences, and examples of this are given below (see section 6.7). These also have implications with respect to the effects of gaseous pollutants, especially through acid rain (see section 9.1). However, the mechanisms involved are obscure and pH differences are often related also to the buffering capacity of the substratum or nitrogenous enrichment; substrata subject to such enrichment are *hypertrophicated* and often support communities restricted to such habitats (e.g. *Xanthorion parietinae*; see section 6.7).

The mineral content of the substratum can also be of major importance. As in vascular plants, the presence of calcium appears to be particularly critical, so much so that separate series of lichen communities develop on *calcareous* (calcium-containing) substrata such as limestones and concrete, and on *siliceous* (silicates predominating) ones. The effect of most minerals is uncertain but several lichens favour rocks rich in heavy metals, for example *Acarospora sinopica*, *Lecanora epanora*, *Rhizocarpon oederi* and *Stereocaulon pileatum* (see sections 5.8, 9.2).

6.2   Limiting environmental factors

The lichens able to establish and grow on a substratum are restricted not only by its nature but by a variety of environmental parameters. These include temperature regimes, wetting and drying frequency, humidity, illumination, and pollutants—factors which also affect vascular plants and bryophytes. Examples of the importance of environmental factors in lichen ecology are given below and in chapters 4, 5, 7 and 9.

6.3   Lichen–substratum interfaces

In the case of lichens on bark or wood those fruticose species of *Letharia* and *Usnea* which have persistent attachment points (*holdfasts*) produce rope-like bundles of hyphae which force their way between the periderm cells (figure 6.1*A*) and can even reach just beyond the cambium in rare cases. The penetrating hyphae may be derived from either the cortex (*Ramalina*) or axis (*Usnea*). In the case of foliose species, the degree of penetration by rhizines is much less and usually restricted to the outermost layers of dead bark cells which are scarcely disrupted (figure 6.1*B*), and the same is true of crustose lichens with a superficial thallus.

The endophloeodal lichens, in which the thallus develops under the surface layers of the bark, although more intimately associated with their host, are also normally confined to dead tissues although these may be more disrupted (figure 6.1*C*); bark cells can also be used in producing composite tissues which are an integral part of the ascomata. It has been suggested that lichens with such habits may obtain some nutrients from the bark but there is no experimental work to support that thesis. Studies with fruticose species failed to identify any such effect.

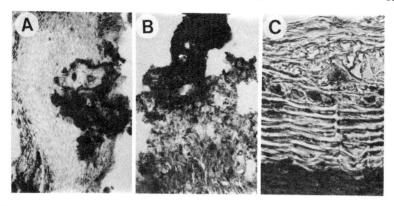

Figure 6.1 Bark penetration by corticolous lichens of different habit. *A, Letharia vulpina* (fruticose) holdfast, × 50. *B, Parmelia sulcata* (foliose) rhizine, × 125. *C, Graphina anguina* (crustose endophloeodal), × 325.

The situation with rock substrata parallels that on bark. Holdfasts and rhizines penetrate in a similar manner and can be expected to have a role in the breaking up of rock surfaces; crustose species penetrate over the whole lower surface, tending to spread horizontally and so inducing a tendency of the surface to flake. Penetration is usually not more than several millimetres but is recorded to a depth of 16 mm. Endolithic lichens have their hyphae and photobiont cells entirely immersed in channels or locules in the rock which they produce themselves. The activities of rock-inhabiting lichens have implications for weathering and pedogenesis.

6.4 Weathering and pedogenesis

In addition to mechanical effects, crustose lichens affect the chemistry of the rocks on which they grow in a series of activities collectively termed *biochemical weathering*. These activities are not primarily due to characteristic lichen products such as depsides and depsidones, which are scarcely soluble in water and have rather limited chelating abilities, but to other readily soluble organic acids secreted by the mycobiont, especially oxalic acid.

Critical studies involving the SEM, microprobe spectra and the culture of both mycobionts and other microfungi (especially Hyphomycetes) have contributed to major clarifications of the situation within the last five years. Oxalic acid is now established as the cause of surface etching, such as pitting, seen in minerals beneath crustose thalli (figure 6.2*B*); in some cases

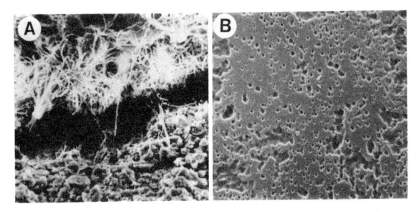

Figure 6.2 *A, Pertusaria corallina* mycobiont-basalt interface, × 1000. *B*, Etched surface of a potassium feldspar crystal due to action of oxalic acid, × 3000. Scanning electron micrographs. *A* from D. Jones, M.J. Wilson and J.M. Tait (1980), *Lichenologist* **12**, 277–289; *B*, from M.J. Wilson and D. Jones (1983), in *Residual Deposits* (R.C.L. Wilson, ed.), Blackwell Scientific Publications, Oxford, 5–12.

this can extend to honeycombing, rendering grains extremely fragile so that they easily disintegrate into finer particles. It is also able to convert some minerals to siliceous relics (for example chrysolite in serpentine to fibrous silica by the removal of magnesium); secondary poorly organized (non-crystalline) weathering products such as silica gel and aluminous or ferruginous compounds arise, and crystallization of oxalates occurs both at the rock-lichen interface and within the thallus itself. Calcium oxalate is the most commonly formed oxalate and its crystals are often the cause of 'pruina' on the thallus surface (see section 2.2). On serpentine, which contains little calcium, magnesium oxalate (the mineral glushinskite) is produced; copper and manganese minerals can be converted to oxalates in the same way.

The importance of biochemical weathering of rocks by lichens in soil formation (*pedogenesis*) is not entirely clear. This is feasible from the mineralogical standpoint as the mineral composition of young soils is often compatible with oxalic acid action on the rocks, but a wide range of Hyphomycetes in soil form oxalic acid and have comparable effects (e.g. *Aspergillus niger*). Citric acid, produced by several soil fungi, for example *Penicillium citrinum*, has also been thought to be significant as it has strong chelating abilities.

Whether lichens actually cause rocks to break down more quickly than ones from which they are absent is a matter for conjecture. Although lichens are active in this regard, the presence of a continuous lichen crust moderates the effects both of chemicals in rain and wind, and especially of blasting by sand or blown ice crystals. In extreme environments, these latter two abrasive factors can scour lichens off rock surfaces and inhibit their colonization. Nevertheless, the depth of weathered material on Hawaiian lava flows was found in one study to be 0.142 mm beneath *Stereocaulon vulcani* and only 0.002 mm on a bare surface of the same age.

Humic materials produced by the breakdown of the basal parts of apically extending fruticose lichens, for example *Cladonia portentosa* and *C. stellaris*, also contribute to soil structure, and these and squamulose lichens adapted to unstable soils (see section 2.4) have some role in binding the surface together so reducing the effects of erosion. The presence of a lichen cover on soil will also facilitate the trapping of wind-blown seeds and so contribute to the establishment of vascular plants.

6.5 Competition and succession

On rocks, soil and often trees, algae, or more commonly cyanobacteria such as *Gloeocapsa* or *Nostoc*, are amongst the first colonizers. On the ground, whether these are followed by lichens, mosses or phanerogams depends on the nature of the habitat and the environment. Lichens may have a successional role in some cases, but are themselves the climatic climax in harsh environments such as tundra lichen heaths. The colonizing species also vary, but appreciable covers can be obtained in a relatively few years in some instances, as after heath fires and on man-made substrata. In dune situations, lichens can only occur *after* the sand has been consolidated to some extent by grasses like *Ammophila arenaria* and represent a seral stage later giving way to shrubby vegetation.

It is often said that crustose lichens are the first colonizers of rocks, but this is not always so. In some cases foliose lichens such as *Parmelia* and *Umbilicaria* species colonize before crustose ones, becoming eliminated as mosaics close.

The slow growth of lichens and the time taken to develop communities hinders studies on succession, but of special interest are the investigations on succession on twigs in Europe by Degelius. In 1964 the examination of twigs from 330 trees of *Fraxinus excelsior*, for example, showed clearly that protococcoid algae and dematiaceous Hyphomycetes were the first colonists; *Physcia tenella* and *Xanthoria polycarpa* appeared in 3- to 5-year-old shoots on leaf scars and other uneven areas, but internode colonization by lichens did not start until 8- to 9-year-old shoots when *Arthopyrenia punctiformis* and *Lecanora carpinea* were the major colonists. A complementary study in 1978 considering other trees confirmed this as a general pattern, emphasizing also *Arthonia punctiformis* as an early internode colonist. Older twigs had a lower density of species, probably owing in this case to the inability of the numerous crustose lichens to compete with the foliose ones, which are faster-growing and can cover them.

When foliose species become senescent and die, often from the centre outwards, the bare areas of substratum made available can be colonized by the same lichen or different foliose or crustose species. It is uncertain what competition (if that is what really takes place) is for in such instances, but space and light may be of key importance.

Complex situations occur in the establishment of two-dimensional *mosaics* of thalli often with a large number of individuals of the same or different species completely covering the substratum. Crustose lichen mosaics are often spectacular (figure 6.3) and may be regarded as permanent in some cases with any growth being in terms of thallus thickness rather than area. The edges of colonies of the same species are often marked by touching photobiont-free prothalline tissue appearing as black or white zones (e.g. *Fuscidea cyathoides*). In the case of sexually-dispersed lichens, each thallus in a mosaic may belong to a different

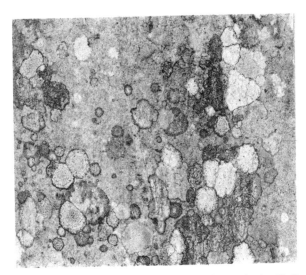

Figure 6.3 Example of a lichen mosaic comprising *Fuscidea cyathoides*, *Huilia tuberculosa* and *Rhizocarpon geographicum* on pre-Cambrian rhyolite tuff in Gwynedd, Wales, showing different types of contact between thalli, $\times \frac{1}{2}$. Photograph A. Pentecost.

genotype when of the same species. The study of lichen mosaics over long periods is fascinating and yields a large amount of information on species–species interactions. For example, one study over 14 years in Wales showed that *Ochrolechia parella* could overgrow *Rhizocarpon obscuratum* but reached an equilibrium on contact with *Lecanora gangaleoides*, whereas *L. gangaleoides* was unable to overgrow colonies of *R. obscuratum* which persisted as islets in its thallus throughout the period.

6.6 The sociological approach

Within a single climatically uniform region, each particular substratum tends to assume, eventually, a characteristic and often remarkably uniform assemblage of lichen species under the influence of similar environmental factors. The occurrence of such noda is the basis of the sociological approach which aims to recognize and characterize them. Communities can be identified by the use of quantitative methods based on randomly collected data sets, or by intuitive ones. A combination of approaches is ideal, but with particularly experienced field workers the intuitive is difficult to surpass.

In order to facilitate communication, systems of naming plant communities have been developed. These have been the subject of extensive work over more than 60 years and a variety of systems have emerged. However, the naming of plant communities was put on a more unified base in 1976 with the publication of a 'Code of Phytosociological Nomenclature' by the International Society for Vegetation Science. A hierarchical system is used of which only two categories need concern us here.

(1) *Alliance*, the name of which ends in the suffix '*-ion*', for example the *Xanthorion parietinae*. Each alliance includes one or more 'associations', one of which is its 'type'.

(2) *Association*, the name of which ends in the suffix '*-etum*', for example the *Physcietum caesiae*. This is the basal unit, equivalent to 'species' in organism taxonomy. Each association must now be typified by a 'relevé' (i.e. a list of species present together with their frequencies). The first type relevé fixes the application of the community name.

The use of the categories 'federation' and 'union', together with special concepts associated with them, is now obsolete but still features in some introductory texts.

Sociological systems are widely used in Europe, and a detailed scheme for the British Isles was published in 1977 (table 6.1). Provided that a

Table 6.1 The principal alliances of lichen communities recognized in the British Isles. Adapted from P.W. James, D.L. Hawksworth and F. Rose (1977) in *Lichen Ecology* (ed. M.R.D. Seaward), Academic Press, London etc, 295–413.

Corticolous communities

Calicion hyperelli (9)[1]	*Parmelion laevigatae* (1)
Cladonion coniocreae (2)	*Parmelion perlatae* (1)
Graphidion scriptae (4)	*Pseudevernion furfuraceae* (2)
Lecanorion subfuscae (1)	*Usneion barbatae* (5)
Lecanorion variae (4)	*Xanthorion parietinae* (8)
Lobarion pulmonariae (1)	

Limestone communities

Aspicilion calcareae (5)	*Xanthorion parietinae* (2)

Siliceous rock communities

Acarosporion sinopicae (4)	*Pseudevernion furfuraceae*
Lecideion tumidae (4)	*Rhizocarpion alpicolae* (3)
Leprarion chlorinae (16)	*Umbilicarion cylindricae* (2)
Parmelion conspersae (3)	*Xanthorion parietinae* (1)

Marine and maritime communities
  Four associations[2]
Terricolous communities
  About seven associations[2]
Aquatic communities
  Four associations[2]

[1] Numbers in brackets are the number of named associations accepted in each alliance.
[2] Associations not referred to an alliance.

'broad-brush' approach is adopted, this shorthand system has a great deal to commend it in floristic and ecological studies.

Lichen-dominated ecosystems, such as those on e.g. a leaf surface or bark crevice, are just as suitable for the use of sociological systems as the communities that include the trees themselves. Of course, in the case of terricolous desert and tundra situations lichens dominate some systems; there is thus a wide continuum of scale.

The study of lichen communities is still in its infancy. The examples given in the following sections are therefore necessarily drawn largely from Europe.

6.7 Corticolous communities

Lichen communities developed on bark, *corticolous communities*, are spectacular when they comprise large pendent fruticose or foliose species. This is especially so in the boreal coniferous zone, where the low bark pH of spruces, pines and birches favours the development of the *Pseudevernion furfuraceae*, which includes species of *Alectoria, Bryoria, Evernia,*

Hypogymnia, Parmelia (e.g. *P. saxatilis*), *Pseudevernia* and *Usnea*. Lichen-laden trees are cut by Lapps and Eskimos to provide winter fodder for reindeer when the ground is frozen. The lichens can also harbour major economic forest pests, for example the Hemlock Looper (*Lambdina fiscellaria*) in Canada. On twigs a colourful assemblage of yellow (*Cetraria pinastri*) and brown (*C. sepincola*) *Cetraria* species and the yellow-green. *Parmeliopsis ambigua* is often conspicuous; on the bases it marks the mean late snow level. In the British Isles the boreal coniferous communities are largely confined to the Scottish Highlands.

In temperate forests the deciduous trees generally have a pH value higher than 5, while those of the coniferous forests tend to have a lower one. The main community in lowland parts of western Europe is the *Parmelion perlatae* comprising a wide range of species including *Parmelia caperata, P. perlata, Pertusaria albescens* and *P. pertusa*. However, each tree provides a wide range of ecological niches where different communities can develop. For example, on nutrient-enriched bark *Xanthorion parietinae* (including *Physcia, Physconia, Ramalina* and *Xanthoria* species) develops, and on old dry bark not subject to direct rain the *Lecanactidetum premneae* (including *Lecanactis lyncea, L. premnea* and *Schismatomma decolorans*). The climax community on such bark in Europe is the *Lobarion pulmonariae* (including *Lobaria, Nephroma, Pannaria, Parmeliella, Sticta* and *Thelotrema* species) but this is now scarce in many areas (see figure 7.6); in montane European forests this community also may occur on *Abies* which has a higher bark pH than most other conifers. Exceptionally high rainfall (over $127\,\mathrm{cm\,yr}^{-1}$) results in leached bark with a pH below about 4.6 when a special climax develops, the *Parmelion laevigatae* (including *Cetrelia olivetorum, Mycoblastus* and *Ochrolechia* species, *Parmelia laevigata* and *P. taylorensis*). Conversely, particularly base-rich bark can support the now scarce *Teloschistetum flavicantis*.

The smooth bark of twigs of deciduous trees in the British Isles may be colonized by a mixture of *Arthonia* and pyrenocarpous species (*Arthopyrenietum punctiformis*), a mosaic of crustose lichens with superficial thalli (*Lecanoretum subfuscae*), and shaded older smooth bark by the *Graphidetum sciptae* (including *Graphina, Graphis, Opegrapha* and *Phaeographis* species) or *Pyrenuletum nitidae* (including *Enterographa crassa* and *Arthonia, Opegrapha* and *Pyrenula* species). In the tropics, rather similar very species-rich communities make up mosaics on a wide range of trees, including members of the Graphidaceae, Pyrenulaceae and Thelotremataceae. Leaf-inhabiting (*foliicolous*) lichens are a special feature of the tropics (see figure 2.1).

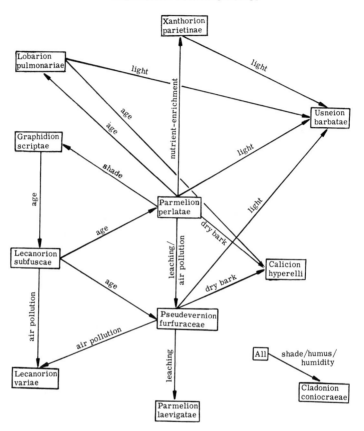

Figure 6.4 Relationships between the main lichen corticolous communities in the British Isles. Adapted from P.W. James, D.L. Hawksworth and F. Rose (1977), in *Lichen Ecology* (M.R.D. Seaward, ed.), Academic Press, London and New York, 295–413.

The relationship between the main lichen communities on trees in the British Isles is indicated in figure 6.4. The effect of parameters limiting them can often be seen in the field, for example vertical replacement along a felled tree from a wood related to light regimes, differences in humidity and nutrient-enrichment of the bark.

Under pollution stress, communities can be substantially modified. Increasing bark acidity leads to the development of the *Pseudevernion furfuraceae*, not the *Parmelion perlatae*, in areas where it would not otherwise occur, or, by the elimination of competitors, to the *Lecanorion variae* characterized by almost pure swards of *Lecanora conizaeoides*.

6.8 Lignicolous communities

Communities occurring directly on wood (*lignum*), *lignicolous communities*, are often identical to those on bark, but some special associations are also to be found, particularly in the boreal zone. Of especial note are those dominated by Caliciales including many species of *Calicium*, *Chaenotheca* and *Cyphelium*. On worked timber and decorticate (bark-free) areas of trees in areas affected by gaseous pollutants, a community characterized by *Hypocenomyce scalaris* (a lichen also common on burnt stumps) often develops.

Wood which develops a rich lichen vegetation on its surface often appears to show enhanced resistance to wood-rotting fungi. Preliminary studies in Finland have shown that lichen extracts from several species common on fence posts (e.g. *Hypogymnia physodes*) are able to inhibit a wide range of wood-rotting fungi (e.g. *Piptoporus betulinus*) in pure culture.

6.9 Saxicolous communities

Lichen communities developed on rocky substrata, *saxicolous communities*, vary markedly according to the rock type (see section 6.1). In parallel to the case in corticolous lichens, a single rock outcrop can also provide a variety of habitats developing different assemblages of species. This can be especially marked in those on hard limestones. The *Caloplacetum heppianae* (including *Aspicilia calcarea*, placodioid *Caloplaca* species, *Verrucaria* species and other pyrenocarpous lichens on well-lit areas); the *Dirinetum stenhammariae* (dominated by *Dirina repanda* f. *stenhammari*) on dry moderately-lit areas; the *Leproplacetum chrysodetae* (with *Lepraria* and *Leproplaca* species) in shaded crevices with a high humidity; the *Gyalectetum jenensis* (including *Acrocordia conoidea*, *Caloplaca cirrochroa*, *Gyalecta jenensis*, *Porina linearis* and *Petractis clausa*) in shaded moist sites; the *Placynthietum nigri* (including *Collema auriculatum*, *Dermatocarpon miniatum*, *Leptogium lichenoides*, *Placynthium nigrum*, *Thelidium papulare* and *Verrucaria caerulea*) on damp well-lit areas; *Lempholemma* species in hollows where water is temporarily present; and the *Physcietum caesiae* (including *Physcia adscendens*, *P. caesia*, *Phaeophyscia orbicularis*, *Physconia grisea* and *Xanthoria parietina*) where nutrient enrichment occurs, for example on bird perches.

Parallels with the above occur on siliceous rocks, for example the *Cystocoleus–Lepraria–Racodium* community on vertical shaded rocks, and the *Chrysothrix chlorina* and *Psilolechia lucida* associations of dry upland recesses. In exposed situations mosaics of mainly crustose species occur in the *Lecideion tumidae* and the *Rhizocarpion alpicolae* (this last including *Rhizocarpon alpicola* which can tolerate being buried in snow for 2–3 months each year), but striking macrolichen-dominated associations can occur including the widespread *Parmelietum omphalodis* and the more upland *Umbilicarietum cylindricae*.

Bird perches on siliceous rocks may develop associations of the *Xanthorion parietinae*, but in less extreme cases of nutrient enrichment special ones arise, for example the *Candelarielletum corallizae* or *Parmelietum glomelliferae*; the latter includes *Parmelia conspersa, P. glabratula* and *P. loxodes*.

6.10 Marine and maritime communities

Lichen communities on siliceous rocky shores develop as a series of distinctive bands related to the degree of influence of the sea and can be integrated into the general biological scheme for rocky shores used in the British Isles (table 6.2).

Arthopyrenia halodytes and *Lichina pygmaea* sometimes range down into the eulittoral, but it is in the littoral fringe that lichens assume major importance, in the *Verrucarietum maurae*. This blackish zone is rich in species of *Verrucaria, V. mucosa* and *V. halizoa* in the lower part, then *V. striatula, V. amphibia* and lastly *V. maura. Lichina pygmaea* forms a distinct band amongst the uppermost barnacles. Only one lichen, the antarctic *V. serpuloides*, is known to apparently tolerate continuous submersion. In the

Table 6.2. Fletcher's (*Lichenologist* 5, 368–422, 1973) terminology of seashore lichen zones in the British Isles in relation to the general biological scheme of Lewis (*The Ecology of Rocky Shores*, 1964) and named associations.

Zone	Subzone	Lichen community
Terrestrial region	halophobic	e.g. *Parmelietum omphalodis*
	halophilic	e.g. *Parmelietum glomelliferae*
Supralittoral zone*	xeric	*Ramalinetum scopularis*
	submesic	*Caloplacetum marinae*
		(+ Xanthoria)
	mesic	*Caloplacetum marinae*
Littoral zone*	littoral fringe*	*Verrucarietum maurae*
	eulittoral*	Absent
Sublittoral zone*		Absent

*Terms used in the scheme of Lewis (1964).

lower parts of the littoral fringe zone 40–50% submersion is usual, falling to around 1% at its upper limits.

The orange *Caloplacetum marinae* zone is characteristic of the mesic supralittoral, and predominant species include *Caloplaca marina, C. microthallina, C. thallincola, Catillaria chalybeia, Lecanora actophila* and *L. helicopis*. In the upper parts *Xanthoria parietina* and *Physcia tenella* often also occur, making up the submesic supralittoral zone (the lowest part of the shore with foliose lichens).

The predominantly grey *Ramalinetum scopularis* zone is very rich in species and includes *Anaptychia fusca, Lecanora gangaleoides, Parmelia pulla, Lecidea sulphurea, Lecidella subincongrua, Ochrolechia parella* and particularly *Ramalina siliquosa*. Dry shaded recesses in this zone also have a distinctive community, the *Sclerophytetum circumscriptae* with, for example, *Arthonia lobata, Caloplaca littorea, Lecanora praepostera, L. tenera, Roccella* species, and *Sclerophyton circumscriptum*.

The terrestrial zone contains predominantly inland species. The salt-tolerant halophilic section comprises species such as *Caloplaca ferruginea, Parmelia saxatilis* and *Ramalina subfarinacea*, while the halophobic includes species only occurring where they are sheltered from direct sea spray such as *Parmelia omphalodes* and *Sphaerophorus globosus*.

These zones are best developed on hard acidic rocks, and not on limestone, chalk, or friable rocks. On exposed shores increased wave and spray action leads to the raising of heights of zones compared to adjacent sheltered shores. The precise factors causing the zonation patterns to develop are uncertain, but wetting frequency may well be more important than salinity.

Oil spills can cause discolorations, but seem less toxic to lichens than oil dispersal and cleansing agents used in 'clearing up' operations which can cause species to brown and peel away from the rock. However, if removed, significant colonization of at least the commoner species occurs within 5–10 years.

6.11 Aquatic communities

The zonation seen on the seashore is paralleled on hard siliceous rocks in freshwater, especially oligotrophic and nutrient-poor streams, lake margins, and pools. Santesson was able to distinguish four zones in lakes in Sweden and relate the occurrence of some species to the frequency of inundation. For example, the lower *Verrucaria* zone was submerged 80–95% of the time, the upper limit of a *Hymenelia lacustris* zone 25%, while most foliose lichens were absent if there was any total submersion—1–2 weeks of flooding killed the terrestrial species which found re-colonization difficult.

Four aquatic associations have been recognized in the British Isles. These include species such as *Bacidia inundata, Dermatocarpon weberi, Ephebe lanata, Hymenelia lacustris, Placynthium flabellosum, Staurothele fissa, Verrucaria aquatilis, V. hydrela* and *V. kernstockii.*

This phenomenon is widespread and documented not only in Europe but also from India, Zimbabwe and Australia, although different species are often involved. In the latter case lichen zones have proved of particular value in the determination of the water capacity of river channels, and pilot studies indicate they can be used in a similar way in the British Isles.

6.12   Terricolous communities

Lichens growing on the ground, *terricolous communities*, are a particularly important and often dominant component of the ground vegetation in extreme environments, particularly in the tundra but additionally in montane situations. The circumpolar lichen heaths are characterized by tufted fruticose species such as *Alectoria nigricans, A. ochroleuca, Bryoria nitidula, Cetraria islandica, C. nivalis, Cladonia mitis, C. rangiferina, C. stellaris* and *Coelocaulon divergens*, and extend south to play a major part in the vegetation below coniferous boreal forests. The ecophysiology of these communities, which is of particular interest, has already been considered (section 5.9). Reindeer and caribou can consume 3–5 kg of lichens each day in the winter months when they often make up over 50% of the diet; corticolous species as well as terricolous ones may be utilized (see section 6.7). Lichens are low in protein but can make significant contributions to the carbohydrate requirements of caribou.

Some of the same lichens are to be found in montane heathlands, as in the Scottish Highlands where they are also often associated with *Rhacomitrium lanuginosum, Salix herbacea* or ericaceous shrubs. On more lowland heaths, many other *Cladonia* species (e.g. *C. floerkeana*) become important and crustose lichens such as *Baeomyces rufus, Lecidea granulosa* and *L. uliginosa* are frequent. Blanket peat, which is often more continuously moist, often also supports *Cladonia strepsilis, C. sulphurina, Icmadophila ericetorum* and *Pycnothelia papillaria*.

Soils associated with basic rocks have quite a different flora including, for example, *Bacidia sabuletorum, Cladonia pocillum, C. rangiformis, Collema tenax* and *Toninia caeruleonigricans*. If the soils are particularly calcareous and with a pH of about 8, *Fulgensia fulgens, Psora decipiens,* or

Squamarina cartilaginea may occur. Squamulose species with well-developed attachment systems (see section 2.4) are characteristic. Communities developed on coastal sand dunes are often similar but additional species may assume major importance, for example *C. foliacea*.

Many of the species to be found on compacted sandy soils in the British Isles are widespread in hot arid and semi-arid desert regions of the world, for example *Dermatocarpon lachneum, Fulgensia fulgens, Psora decipiens, Squamarina lentigera,* and *Toninia caeruleonigricans.* Additional genera are also characteristic of such habitats, for example *Gonohymenia, Heppia,* and in Australasia the remarkable monotypic *Chondropsis viridis* which curls up into a tight ball when dry but uncurls flat on the surface when wet.

6.13   Man-made substrata

A large number of substrata introduced by humans can be colonized by lichens, including bone (calcicolous species), leather, hair and wool, silk, worked timber, thatch, polystyrene, glass fibre, walls, memorials and sculptures, paint, asbestos-cement, concrete, asphalt, glass, worked iron, and some non-ferrous metals. The use of worked timber and building materials in areas otherwise devoid of woodland and of natural rock outcrops can extend the distributional range of species substantially.

In general the communities developed are species-poor variants of ones also developed on naturally occurring substrata, but some lichens have proved particularly adept at colonizing those introduced by man. Examples include *Caloplaca citrina, Lecania erysibe, Lecanora conizaeoides, L. dispersa, L. muralis, Scoliciosporum umbrinum, Xanthoria elegans* and *X. parietina. L. dispersa* is one of the first lichens to appear on cement-containing substrata, even though it reproduces by ascospores and has *Pseudotrebouxia* (see section 1.6) as photobiont, and gives high percentage cover remarkably rapidly by the formation of numerous individual thalli arising from separate propagules. In conjunction with *Candelariella aurella,* 70–90% cover can be attained in 3–5 years even in the centres of major cities such as London.

A few taxa may even be confined to man-made habitats, for example *Lecanora vinetorum* known from poles in Italian vineyards subjected to spraying with copper fungicides.

6.14   Biodeterioration

The effects of lichens on man-made materials are poorly investigated. Of major concern has been their growth on sculptures which can lead to actual disintegration of the substratum (see section 6.4) but it is also of concern because of the disfigurement produced. On concrete in urban areas, black stains due to colonies of *Candelariella aurella* and *Lecanora dispersa* (see above) which involve particles of soot in their thalli, may be unsightly. In the case of cement-containing materials, there are indications that cement paste rather than included aggregates is affected, and in brick and tiles the clay minerals present appear to be preferentially attacked.

Growths can be removed with a variety of fungicides, for example Bordeaux Mixture (a complex copper hydroxide precipitate), but for use on stone, chemicals lacking copper are to be preferred as greenish stains may remain. Kerosene sprays, diesel distillates and caustic washes have been tried but because of their hazards to users a 4% solution of proprietary borate compound 'Polybor' has much to commend it and is used by the Commonwealth War Graves Commission. Where delicate sculpturing is involved, if sacking is tied over the stone for 4–8 weeks under humid conditions this can also be effective. After death, superficial lichens can be brushed off.

The development of a colourful lichen community can often be encouraged by spraying new building materials with a dilute manure extract which will promote the development of *Xanthorion* communities. This is sometimes required to help new structures tone in to their surroundings relatively quickly.

6.15 Substratum amplitude

The extent to which species are confined to a particular substratum varies enormously. Some lichens have an extremely wide ecological amplitude, for example *Lecanora dispersa*, *Parmelia sulcata* and *Xanthoria parietina*, and even associations can occur on more than a single substratum (e.g. *Physcietum ascendentis*). However, most are restricted to varying degrees and examples may be found in preceding sections.

Most corticolous lichens appear to be bark-type rather than host tree (*phorophyte*) specific, but there are exceptions, as with *Lecanora populicola* on bark of *Populus* species, and a considerable number of often rather doubtfully lichenized pyrenocarpous lichens such as *Arthopyrenia laburni* on *Laburnum* twigs. Foliicolous lichens might be expected to be more restrictive but there is little evidence for this; *Strigula elegans* for example is known from the leaves of at least 99 genera of trees belonging to 51 families.

Conversely, individual types of tree can support a varied range of lichens. In 1974 Rose reported that in the British Isles oak could support 303 lichens, ash 230, elms 171, alder 72 and hornbeam 42. In an unpolluted area, a single ancient oak tree may have as many as 50 different species present on it.

In some cases a species may change from its usual substratum to another; such *substratum switches* can arise for several reasons, the commonest of which is a change in the substratum itself so that it comes to resemble that on which the lichen is usually found. Examples are the usually saxicolous *Lasallia papulosa* and *Parmelia taractica* on conifer wood in Ontario, *Fuscidea cyathoides* on leached bark in high-rainfall parts of the British Isles, and *Physcia caesia* on ash bark in the English Peak District so heavily impregnated with lime dust that it recalls limestone. Of particular interest are species which have been able to switch to man-made substrata, such as *Trapelia coarctata* on rubber dustbin lids and *Xanthoria elegans* on concrete (figure 7.6). *Lecanora muralis* has switched from bird-perching sites to a range of man-made substrata including asbestos cement, asphalt and brick over large often very polluted areas of lowland England; in this case ecotypic genetic changes may also be involved as the species remains confined to its original habit in North America.

6.16 Ecology and speciation

In some cases very similar species have come to occupy separate habitats, indicating that ecological separation can be an important factor in lichen speciation. The increased amplitude of secondary species has already been reviewed (see section 3.8), but there are also examples not involving changes in reproductive strategy.

Parmelia britannica on sunny coastal siliceous rocks differs from the more ubiquitous *P. revoluta* in the dark isidia, slightly narrower lobes and other minor features. Where *Cladonia polycarpia* (containing norstictic and stictic acids) and *C. polycarpoides* (norstictic alone) are sympatric in North America, Culberson found that these occurred on sandy and clay soils respectively. *Parmelia caperata* primarily occurs in trees in North America, while the very closely related but chemically distinct *P. baltimorensis* is essentially saxicolous; it has been suggested that *P. caperata* may have sustained genetic impoverishment through the loss of gannodemes adapted to rock habitats as a result of intense competition from *P. baltimorensis*. Similarly the crustose *Haematomma ochroleucum* var. *ochroleucum* (with usnic acid) is essentially saxicolous in Europe, whereas var. *porphyrium* (lacking usnic acid) occurs on both rocks and trees.

In other cases it is possible that similar species on different substrata currently maintained as separate may not prove to be so if examined critically. For example, is the saxicolous *Opegrapha lithyrga* really distinct from the corticolous *O. vulgata*?

6.17 Interactions with other organisms

Lichens interact with other organisms in a variety of ways, as sources of food, shelter or camouflage. Associations with invertebrates are particularly varied. In damp situations protozoans, nematodes (including, for example, *Aphelenchoides parietinus* from *Xanthoria parietina*), rotifers, oligochaetes and tardigrades may be found, but interactions are more frequent with terrestrial rather than aquatic organisms—especially arthropods.

Lichen-feeders can occur in large numbers in lichen populations and include springtails (e.g. *Hypogastura packardii* on *Parmelia baltimorensis*), psocids (e.g. *Elipsocus mclachani* on *Lecanora conizaeoides* ascomata and *Reuterella helvimacula* on its thallus), oribatid mites (e.g. *Mycobates parmeliae* on *Xanthoria parietina, Scapheremaeus petrophagus* on *Thelidium auruntii*), orthopterans (e.g. *Scirtetica ritensis* on USA desert rock lichens), and molluscs (e.g. *Lehmannia marginata* on *Hypogymnia physodes* and *Lobaria pulmonaria*). Slugs are major grazers of foliose lichens. The cortex and photobiont layer tend to be eaten, while the medulla is left behind.

Lichen mimics include orthopterans, particularly in the tropics, and even amphibians (e.g. the frog *Hyla versicolor* in N. America) and reptiles (e.g. the gecko *Uroplates fimbriata*) but are most frequent in the Lepidoptera. About 30 European moths eat or mimic lichens. Mimicry may be at the larval stage as in the caterpillars of *Cleodares lichenaria* which resemble *Hypogymnia physodes* thalli, or in the adults as with *Cryphaea muralis* and *Ochrolechia parella*. The changing frequencies of the camouflaged and dark (*melanic*) morphs of *Biston betularia* respond to the elimination of lichens by air pollution, and a parallel situation occurs in some other insects such as the psocid *Mesopsocus unipunctatus*. Foliose and crustose lichens can grow on the backs of a group of beetles in New Guinea, including *Gymnopholus lichenifera*, and have also been found on the giant tortoise *Geochelone elephantopus* in the Galapagos Islands.

About 45 species of North American birds make nests partly or mainly of lichens, and examples are also known in Europe and the tropics. In addition to reindeer and caribou (see section 6.12), a wide range of ungulates can utilize lichens as a part of their diet, and some may even be eaten by humans (e.g. *Bryoria fremontii* 'pancakes' by British Columbia Indians, *Umbilicaria esculenta* 'iwatake' in Japan).

Lichen extracts can inhibit the germination and growth of some grasses and fungi (see section 8.6), and lichen thalli (e.g. *Ochrolechia androgyna, Pertusaria albescens*) may overgrow and destroy bryophytes.

CHAPTER SEVEN

BIOGEOGRAPHY

The study of lichen distribution has been held back by inadequate collecting and taxonomy, but progress with respect to particular regions and groups has changed the position materially during the last 15 years.

7.1 World

Lichen distribution patterns follow major climatic and vegetation zones, but families and genera tend to have much wider ranges than is the case in, for example, flowering plants. In Australia most of the important components of the higher plant vegetation belong to genera and families unknown outside the continent. In contrast the main lichens not only belong to genera familiar to botanists from the Northern Hemisphere but in many cases are represented by species very close to those found on the other side of the world. The reason for this is their antiquity and the effects of plate tectonics (see below); cases made for long-distance dispersal rarely seem credible and this is not now thought to be a primary factor, although it certainly arises occasionally (see section 7.7).

Many lichens have a pantropical distribution, as in the corticolous *Coccocarpia pellita* (figure 7.1) and foliicolous *Strigula elegans*. This pattern varies considerably in degree, and a few essentially pantropical taxa such as *Chrysothrix candelaris* extend well into Europe. In some instances widely distributed genera show their optimal speciation in the tropics, for example in *Graphina*, *Pyrenula* and *Thelotrema*.

Particularly remarkable are the 'bipolar' patterns where species occur in the arctic, antarctic, and sometimes a few intervening montane areas, as in *Alectoria nigricans* (figure 7.1), *Cetraria delisei* and *Pseudephebe minuscula*. At least 300 lichens are either circumpolar or circumboreal in the Northern Hemisphere spreading over wide areas of tundra and taiga after the retreats of the Pleistocene ice sheets (figure 7.2). These patterns may be incomplete and a widespread variant is the amphi-Beringian where the distribution is continuous through North America and Asia but the species are absent from Europe (figure 7.2). This is considered to be due to

102

solid ice sheets having eliminated species from much of Europe and east North America but leaving colonies in the less intensely glaciated parts of Alaska and Siberia from which colonization has progressed east and west. In the case of some boreal forest macrolichens, present disjunctions may also be related to ice sheet movements, for example the absence of *Alectoria sarmentosa* from large areas of central Canada and the separation of the arctic *Bryoria subdivergens* and *Rhizocarpon bolanderi* between Greenland, the mountains of Montana and Norway.

Where populations have become isolated geographically or ecologically differentiation of subspecies may occur. In *Cetraria islandica* four are recognized: subsp. *islandica* (circumpolar in the Northern Hemisphere), subsp. *crispiformis* (as subsp. *islandica* but absent across Northern Asia), subsp. *orientalis* (eastern Asia) and subsp. *antarctica* (extreme South America and Australasia).

7.2 Plate tectonics

Independent studies relating present distribution patterns to plate tectonics have indicated the enormous age of some lichen groups and aided the interpretation of current world distributions. In the Opegraphales, Tehler has found by the comparison of cladograms that *Dirina* almost certainly arose before the Late Cretaceous with the ancestral species distributed along the coasts of the Tethys sea, while the precursors of *Roccellina* were established and distributed *before* the break-up of Pacifica about 225 million years ago. This implies that this group is almost twice as old as that of the first angiosperms. Some lichens may be 'living fossils'. The antiquity of the life style is perhaps further substantiated by the discovery in 1975 of certain lichen-like structures in the pre-Cambrian Witwatersrand carbon deposits in South Africa, although the interpretation of these is not yet entirely clear.

Support for antiquity has also now been found in the largely tropical family Megalosporaceae (Lecanorales). Sipman has recently reconstructed a probable evolutionary history relating the development of spore types in this family to the break-up of continents. This family appears to have originated in Palaeozoic times on the eastern part of the Gondwanaland Shield when this included the tropics about 200 million years ago, additional spore types developing in humid coastal forests by around 150 million years ago. Selected spore types later migrated over large areas in the Cretaceous around 100 million years ago, with more limited migrations in subsequent periods. The crustose Roccellaceae and Megalosporaceae

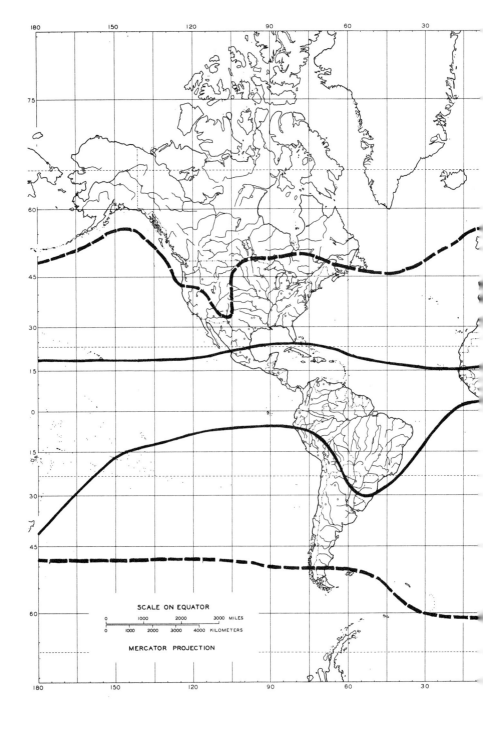

SCALE ON EQUATOR

0   1000   2000   3000 MILES
0 1000 2000 3000 4000 KILOMETERS

MERCATOR PROJECTION

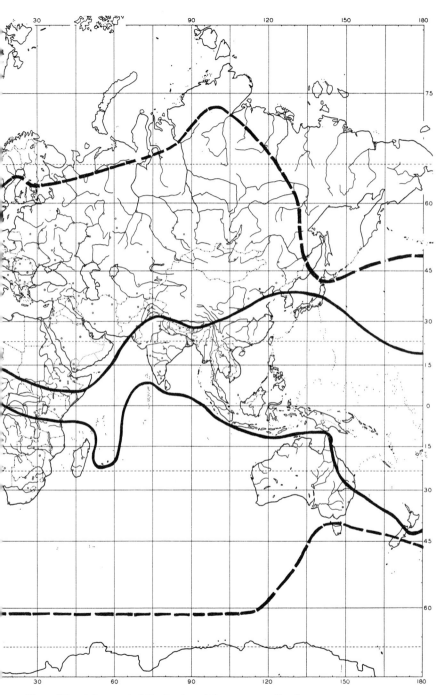

Figure 7.1 *A*, Pantropical distribution of *Coccocarpia pellita* (between continuous lines). *B*, Bipolar distribution of *Alectoria nigricans* (north of the upper and south of the lower broken lines). *A* based on data from L. Arvidsson (1983), *Opera bot.* **67**, 1–96; *B* on data from D.L. Hawksworth (1972), *Lichenologist* **5**, 181–261.

patterns may well be expected to have occurred in numerous other lichen families.

In the case of *Pseudocyphellaria*, Galloway has suggested that the ancestors of present species arose when South America and Australasia (especially New Zealand) were joined, and that after separation of the continents parallel speciation gave rise to related vicariant species occupying similar ecological niches. *Degelia gayana* remains in both Australasia and Chile, with additional species in Australasia and Hawaii.

Studies by Sheard on *Dimelaena oreina* and *Thamnolia vermicularis* and their chemotypes indicate that both genera date from Permo-Triassic times

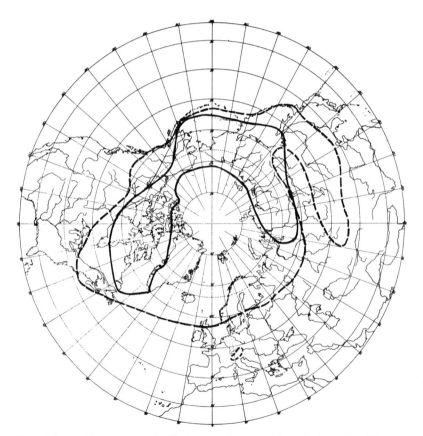

Figure 7.2 *A*, Circumboreal distribution of *Cetraria delisei* (broken line; Northern Hemisphere only). *B*, Amphi-berengian distribution of *C. andrejevii* (continuous line). Based on data from I. Kärnefelt (1981), *Opera bot.* **46**, 1–150.

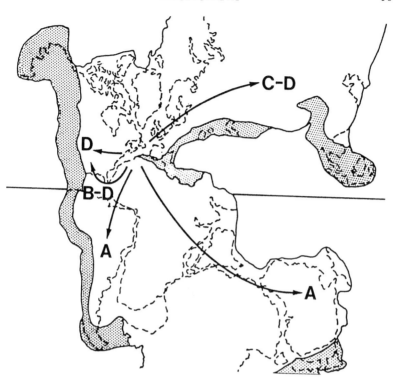

Figure 7.3 Radiation of *Dimelaena* species and chemotypes in Triassic times. *A, D. diffractella. B, D. radiata, D. thysanota* and *D. californica* (or their ancestral species). *C, D. oreina* chemotype with stictic acid. *E, D. oreina* chemotypes with either gyrophoric acid or acid-deficient. Stippling represents continental seas. Adapted from J.W. Sheard (1977), *Bryologist* **80**, 100–118.

and he was able to relate the present distribution of the chemotypes to the established sequence of continental movements (figure 7.3).

7.3 Centres of diversity

Some lichen genera have species concentrated in particular regions of the world and a few representatives extending into other regions. This will be related to their history and date of origin (see above), but has not always been studied in detail. Perhaps the commonest pattern is for genera to exhibit their maximum diversity in the Southern Hemisphere. Most of the approximately 200 species of the temperate genus *Pseudocyphellaria* occur in either Australasia or South America, only five being found in Europe.

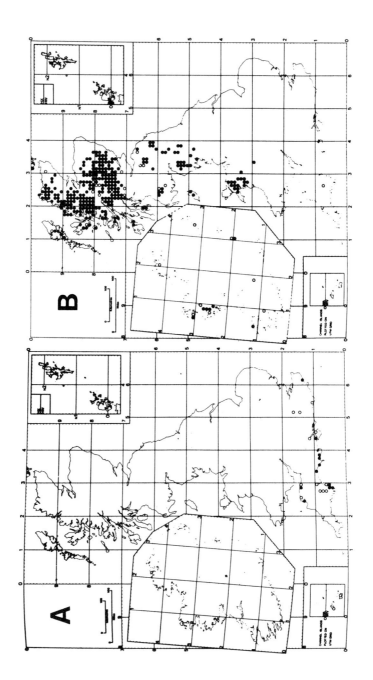

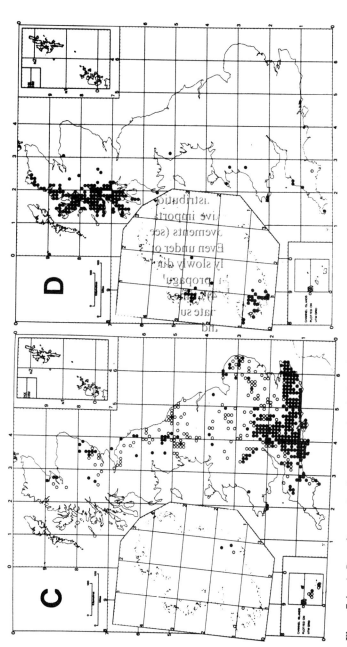

Figure 7.4 A, *Parmelia quercina* (extreme southern). B, *Umbilicaria cylindrica* (northern montane). C, *Anaptychia ciliaris* (eastern). D, *Parmeliella atlantica* (western). o = pre-1960, ● = post-1960. From M.R.D. Seaward and C.J.B. Hitch (1982), *Atlas of the Lichens of the British Isles*, vol. 1, Institute of Terrestrial Ecology, Cambridge.

Similarly, 17–20 species of *Placopsis* occur in South America and Australasia with only two reaching the Northern Hemisphere. The group of *Bryoria* species with lateral spinules (sect. *Divaricatae*) has its centre of diversity in the Himalayas and mountains of China and Indonesia where there are 15 species; four are circumpolar or dissected circumpolar in the Northern Hemisphere, one extends from Asia through the mountainous regions of Europe to the British Isles (*B. smithii*), and two are vicariant endemics in western North America (e.g. *B. cervinula*).

7.4 Factors limiting distribution

Most of the factors limiting the distribution of lichens are shared with the flowering plants but their relative importance varies. In addition to the major factor of continental movements (see above), the rates of dispersal and establishment are critical. Even under optimal conditions lichens are to be expected to spread extremely slowly due to their slow growth rates and time taken to start to form propagules, but once established have exceptional longevity (chapter 4); time is consequently critical.

The availability of an appropriate substratum is also crucial, for example the scarcity of trees in Iceland and Spitsbergen severely limits the proportion of corticolous species compared with more temperate regions.

In view of the balanced physiological relationship which exists between a lichen thallus and its environment (see chapter 5), correlations with isotherms (temperature), and isopleths (rainfall, wet-days, and insolation) undoubtedly occur. Few attempts to establish them have been made, and most of these are on a national scale and of somewhat uncertain value. Examples are *Bryoria furcellata* apparently related to the 13°C isotherm in both Canada and Europe, *Parmelia pastillifera* to both the 2°C January isotherm and 810 mm (32 ins) p.a. isopleth in the British Isles, and *P. soredians* to the mean 4.5 h sunshine day^{-1} isopleth in the British Isles and France.

7.5 European patterns

In Europe, the boreal coniferous forest zone flora is of major importance in Scandinavia (see section 6.7). Some species that do not yet occur throughout that zone, such as *Bryoria furcellata*, may still be migrating from the east. The lichen flora of mountainous areas of central and eastern Europe reflects that of the boreal zone, having many species in common with it, but also lacking some and supporting many additional taxa. In Mediterranean

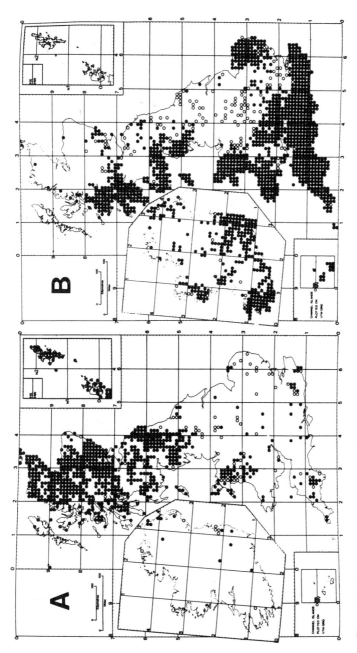

Figure 7.5 *Bryoria fuscescens* (substrata pH *c*. 3.5–5.0). *B*, *Parmelia caperata* (substrate pH *c*. 5.0–6.0). o = pre-1960, ● = post-1960. From M.R.D. Seaward and C.J.B. Hitch (1982), *Atlas of the Lichens of the British Isles*, vol. 1, Institute of Terrestrial Ecology, Cambridge.

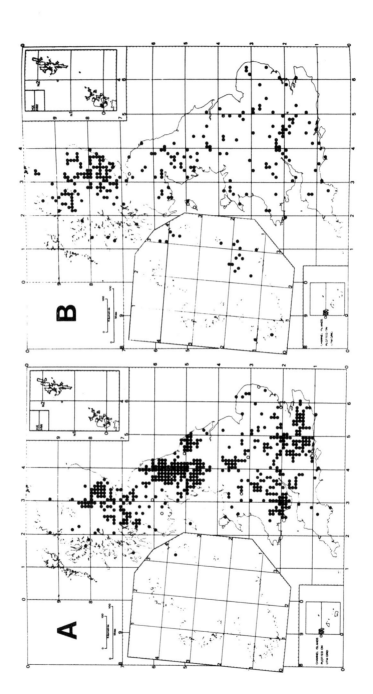

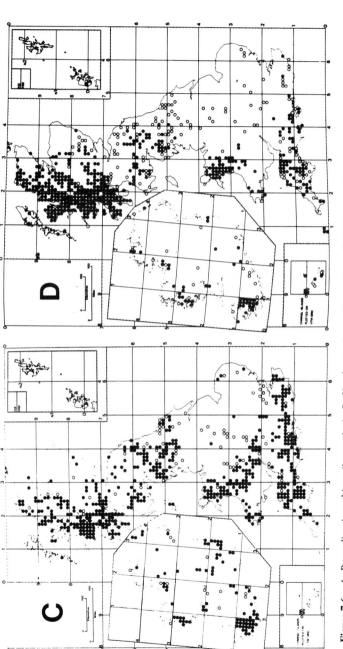

Figure 7.6 *A, Parmeliopsis ambigua* (increasing). *B, Xanthoria elegans* (increasing). *C, Thelotrema lepadinum* (decreasing). *D, Lobaria pulmonaria* (decreasing). o = pre-1960, ● = post-1960. From M.R.D. Seaward and C.J.B. Hitch (1982), *Atlas of the Lichens of the British Isles*, vol. 1, Institute of Terrestrial Ecology, Cambridge.

areas the montane zone recalls the oceanic areas further north and west, including species such as *Parmeliella plumbea.*

The variation in altitude and climate even over short distances in the Alps can be considerable and it is perhaps not surprising that some species appear to be endemic to parts of this region; for example *Buellia subsquamosa* and *Caloplaca microphyllina* from dry valleys of the inner Alps. Such valleys also have strong affinities with the Mediterranean flora, where many genera have representative species widely distributed around the Mediterranean Sea including *Diploschistes ocellatus, Pannaria olivacea* and *Roccella tinctoria.* The communities of the *Xanthorion parietinae* also become much more important around the Mediterranean, perhaps mainly due to nutrient-rich dust being deposited in the summer drought periods.

A major separation in the more northerly European lichen flora is between species in areas with a continental climate (cold winters and hot summers) similar to that on the North American shield, and an oceanic one (mild winters and cooler summers). The tendency to an oceanic distribution varies greatly, some species' distribution being so in a broad sense (e.g. including the whole of the British Isles) such as *Cladonia portentosa,* and others confined, north of the Alps, to the Atlantic seaboard itself as in the case of *Parmeliella atlantica.*

7.6   British patterns

The distribution of lichens within the British Isles has been mapped more intensively than in any other country in the world, with the emphasis being on corticolous species. Even allowing for the effects of man (see below), distinctive patterns can be identified, forming seven groups within the corticolous flora alone. Examples of montane-northern (e.g. *Umbilicaria cylindrica*), extreme southern (e.g. *Parmelia quercina*), eastern (e.g. *Anaptychia ciliaris, Parmelia acetabulum*) and western species (e.g. *Parmeliella atlantica*) are shown in figure 7.4. In some cases distributions are complementary and related to community types which are correlated with factors such as bark pH, as with *Bryoria fuscescens* and *Parmelia caperata* (figure 7.5), whose communities are discussed in section 6.7.

7.7   Effects of man on patterns

The most marked deleterious effects of man are undoubtedly air pollution and the clearance of long-established woodland. The effects of air pollution vary according to the species (see section 9.1), and have eliminated lichens from large areas of Europe and North America, for example *Parmelia caperata* (figure 7.5) which has been lost from much of central England and the northern European plains in the course of the last century. Air pollution can, however, favour some species, as with the small foliose lichen *Parmeliopsis ambigua* (figure 7.6) which has been able to extend from the Scottish pinewoods southwards due to the lowering of the bark pH of deciduous trees by acid rain (see section 9.1), and *Lecanora conizaeoides* probably due to these factors and also the removal of competitors (see section 9.1). Many of the old forest indicator lichens (see section 9.4) are also sensitive to air pollutants as well as woodland clearance, so they declined more dramatically than many others, as with *Lobaria pulmonaria* and *Thelotrema lepadinum* (figure 7.6), the former now appearing western due to these factors (but actually absent in the most extreme oceanic conditions).

The asphalting of roads and advent of cars has led to fewer dust-enriched roadside trees and the decline of *Xanthorion* species such as *Physcia clementei* and *Teloschistes flavicans*. On the other hand, agricultural spraying of nitrogen- and phosphorus-containing fertilizers can encourage the development of some common *Xanthorion* species, such as *Diploicia canescens* and *Xanthoria parietina*, and lead from cars has led to dramatic extensions of the range of *Stereocaulon pileatum* from metal-rich montane rocks to lowland roadside walls. Effects of many agricultural chemicals remain poorly documented, but herbicides can be locally significant, as may water pollution, the burning of grouse moors, quarrying, trampling and even over-collecting (a major factor in Victorian times and still with us).

The importance of man-made substrata has already been discussed (section 6.13). *Xanthoria elegans*, formerly an upland bird-perching site species, now has a wide lowland distribution, mainly on concrete and asbestos-cement (figure 7.6).

Introductions of species by man are also significant. In one 1977 study in west Greenland, 18 of 36 lichens found on timber used for drying cod were considered imported. Other examples include the introduction of *Lecanora conizaeoides* into several towns and cities of North America (e.g. St John's in Newfoundland; Boston, Massachusetts) and even New Zealand (Dunedin), and of *Xanthoria parietina* into urban areas of south-eastern Australia.

CHAPTER EIGHT

SECONDARY METABOLITES

About 550 natural products, including 350 secondary metabolites, are known from lichens, including aliphatic acids, *para*- and *meta*-depsides, depsidones, benzyl esters, dibenzofurans, usnic acids, xanthones, anthraquinones, terpenoids and pulvinic acid derivatives; representative formulae are shown in figure 8.1.

The majority of these compounds are only known from lichens, but it is important to note nevertheless that about 30 'lichen' products are known in other fungi, for example the common depside lecanoric acid in *Pyricularia*, 6-hydroxymethyl-eugenin in *Chaetomium minutum* (also in the lichen *Roccella fuciformis*), and especially physcion (parietin) in species of *Achaetomium, Alternaria, Aspergillus, Dermocybe, Penicillium*, and also the vascular plants *Rheum, Rumex* and *Ventilago*. The list is extending as additional non-lichenized taxa are examined. Further, many fungal products are chemically very similar to lichen metabolites (e.g. caperatic acid to agaricic acid). Some of these may be expected to be identified in lichens.

8.1 Biosynthesis

Two main biosynthetic routes are involved in the production of secondary metabolites in fungi, (1) the shikimic acid pathway leading to products such as phenylalanine and tyrosine, and (2) the acetate–malonate pathway leading to compounds like endocrocin and 6-methylsalicilic acid. These same pathways are the key to the production of most lichen products. In the shikimic acid pathway, the shikimic acid precursor is formed by a series of steps starting with phosphoenol pyruvic acid and erythrose phosphate. The compounds of particular interest arising from this route in lichens are pulvinic acid (tetronic acid) derivatives, many of which are brightly coloured (e.g. bright yellow-green vulpinic acid of *Letharia vulpina*, bright yellow stictaurin in *Candelariella vitellina* and *Pseudocyphellaria aurata*). Terphenylquinones produced by the same route seen in some genera of the

116

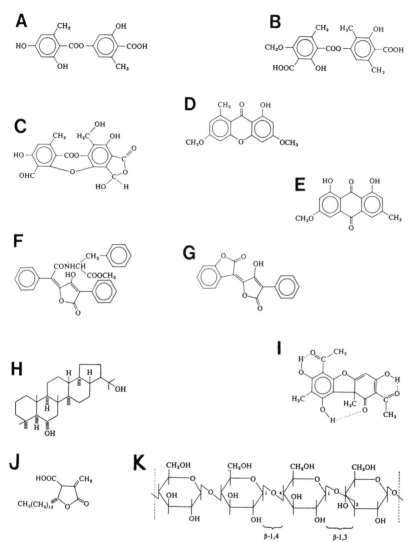

Figure 8.1 Examples of lichen products. *A*, Lecanoric acid (an orcinol series *para*-depside). *B*, Squamatic acid (a *β*-orcinol series *para*-depside). *C*, Salazinic acid (a *β*-orcinol series depsidone). *D*, Lichexanthone (a xanthone). *E*, Physcion (an anthraquinone). *F*, Rhizocarpic acid (a pulvinic acid derivative). *G*, Calycin (a pulvinic acid derivative). *H*, Zeorin (a triterpene). *I*, (−)-usnic acid (an usnic acid). *J*, (+)-protolichesterinic acid (a higher aliphatic acid). *K*, Lichenin (a polysaccharide of mycobiont walls). Formulae from C.F. Culberson (1969), *Chemical and Botanical Guide to Lichen Products*, University of North Carolina Press, Chapel Hill.

Peltigerales are also known in the non-lichenized Aphyllophorales (e.g. polyporic acid).

Most secondary metabolites in lichens are formed by the acetate–malonate pathway, which also gives rise to penicillic acid and many other biologically active compounds in non-lichenized Hyphomycetes. Acetate units condense to form, with the help of malonate, orsellinic acid which is the key to the production of depsides, depsidones, dibenzofurans and higher aliphatic esters. A route excluding malonate leads to a mevalonic acid pathway and the production of triterpenoids and other compounds (figure 8.2).

There are now reports of several lichen products being produced in pure cultures of mycobionts (e.g. squamatic acid from *Cladonia crispata*, salazinic acid from *Ramalina siliquosa*), but there are indications that photobionts may be involved in many cases. In *Lobaria amplissima* and *L. erosa*, lichen products occur in the thallus with green algae but not in the cephalodia with cyanobacteria (see section 2.6). Careful studies on re-synthesized *Cladonia cristatella* by Culberson and Ahmadjian led them to propose in 1980 that photobionts could produce inhibitors of a de-carboxylase converting orsellinic acid to orcinol (a route to many fungal products) so that esterification of orsellinic acid units assumes major importance. Without the photobiont, non-characteristic products, pro-bably including skyrin which is absent in intact thalli, were produced.

Work on the biosynthesis of individual metabolites in lichens using ^{14}C-labelled compounds is meagre, but hypotheses can often be proposed on the basis of what is known of the biosynthesis of similar fungal products.

8.2 Microchemical determination

A variety of methods are utilized in the microchemical determination of the lichen products used in systematics. The simplest are *reagent tests*, the first of which, iodine (I; blue with certain polysaccharides) was used by Nylander in both lichenized and non-lichenized ascomycetes in 1865. In the following year he introduced potassium hydroxide (K; coloured salts with quinoid pigments, some depsides, and many *β*-orcinol depsidones) and bleach (C; pink or red with aromatic compounds with two free hydroxyl groups *meta* to each other). Also used are K followed by C (KC), where the K first hydrolyses ester bonds so opening hydroxyl groups for action with C, and C followed by K (CK), where the method of reaction is unknown. In 1934 the Japanese lichenologist Asahina introduced the use of *para-*

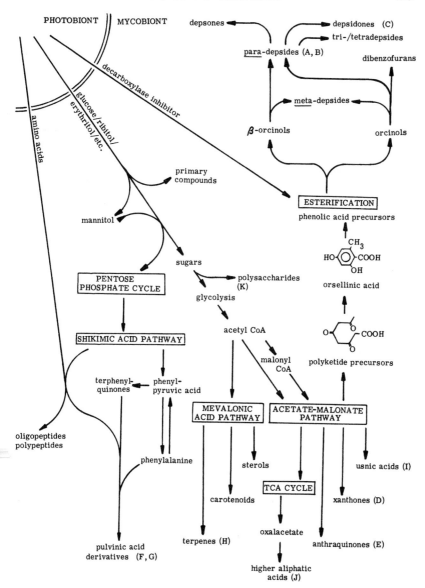

Figure 8.2 Probable pathways leading to the major groups of lichen products. The letters by compound types refer to examples in figure 8.1.

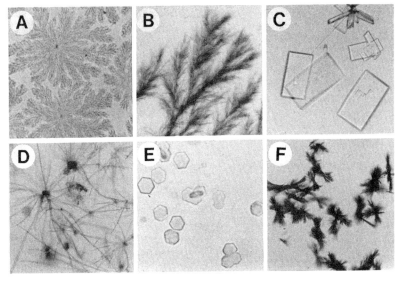

Figure 8.3 Microcrystal tests for lichen products. *A*, Caperatic acid (in GE; glycerol, 1: glacial acetic acid, 3), × 5. *B*, Evernic acid (in GE), × 5. *C*, Barbatic acid (in GE), × 5. *D*, Physodic acid (in GAW; glycerol, 1: ethanol, 1: water, 1), × 5. *E*, Bellidiflorin (in GAAn: glycerol, 2: ethanol, 2: aniline, 1), × 5. *F*, Salazinic acid (in KK; 5% potassium hydroxide, 1:20% potassium carbonate, 1), × 10.

phenylenediamine (P or PD) which reacts with aromatic aldehydes to form coloured Schiff bases.

A specific series of microcrystal tests able to be performed on microscope slides was invented by Asahina in 1936–40 (figure 8.3). Many of these were extremely sensitive, in some cases enabling compounds differing only in single hydroxyl groups to be reliably distinguished. The rapid progress of lichen chemotaxonomy from the 1930s owes much to the development of this method which requires minimal equipment and is well-suited to practical classes.

Further impetus was provided by the introduction of thin-layer chromatography (t.l.c.) first used for lichens in 1963, and especially by the publication of standardized techniques, the first in 1970, which have been rapidly adopted by lichen taxonomists world-wide (figure 8.4). Lichens are one of the few groups where descriptions lacking chemical data would be regarded as incomplete. More sophisticated methods suitable for use only in major chemical laboratories, particularly high performance liquid chromatography (h.p.l.c.), are being increasingly used; h.p.l.c. is re-

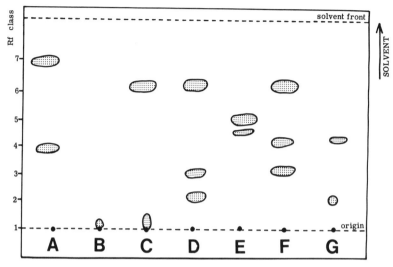

Figure 8.4 Tracing of t.l.c. plate to show Rf classes of some widespread lichen products in TDA (toluene, 180: dioxan, 60: acetic acid, 8; 'solvent A'). *A, Parmelia acetabulum* (norstictic acid, 4; atranorin, 7). *B, Cladonia fimbriata* (fumarprotocetraric acid, 1). *C, Usnea florida* (usnic acid, 6; thamnolic acid, 1). *D, Squamarina cartilaginea* (usnic acid, 6; psoromic acid, 3; conpsoromic acid, 2). *E, Enterographa crassa* (confluentic acid. 5; accessory, 4–5, UV + blue after charring). *F, Parmelia conspersa* (usnic acid, 6; norstictic acid, 4; stictic acid, 3). *G, Cladonia crispata* (barbatic acid, 4; squamatic acid, 2). For details of colours see F.J. Walker and P.W. James (1980), *Bull. Br. Lichen Soc.* **46** (suppl.), 13–29. Plate prepared by F.J. Walker.

markably sensitive and can be used on minute thallus fragments of less than 100 μg. Normal chemical procedures, such as mass spectrometry, are utilized to elucidate structural formulae but for this much larger samples are often needed.

References to articles explaining the procedures used in the microchemical identification of lichen products are included under Further Reading.

8.3 Location and concentration

Secondary metabolites in lichens are mainly deposited as crystals on the surfaces of hyphae. These may be accumulated to appreciable concentrations, in some cases to 10% of the dry weight of the thallus. Studies of field collections of varying age have showed little evidence of variation in the amounts produced; however, recent work on cultured clones of *Cladonia cristatella* revealed that while products of the barbatic acid pathway did not vary in concentration with age, those of the didymic acid pathway were increasingly abundant in older squamules; significant quantitative differences existed between clones, and temperature also affected amounts formed. Light was not important in this instance but does appear to be so in some others (see table 5.7).

The compounds are usually produced in the medulla *or* the cortex, a common pattern being to have one compound in the cortex and different one(s) in the medulla. The common products atranorin and usnic acid are characteristically cortical; physcion and several related compounds in the Teloschistales are deposited on the surface of the cortex. Where the compound is bright-coloured the restriction to particular tissue layers is easily seen, as in the bright orange solorinic acid in the medulla of *Solorina crocea*.

Localization can be more restricted, for example fumarprotocetraric acid in the upper parts of *Cladonia rangiferina* podetia which are still growing and around soralia in *Bryoria lanestris*, the dark red haemaventosin in the apothecia of *Haematomma ventosum*, and norstictic acid in the hymenium of *Letharia columbiana*. Localization may also depend on the photobiont present (see section 2.6).

8.4 Systematic value

Metabolite data is probably used more extensively in routine identification work in lichens than with any other group; physiological and biochemical tests, such as are used regularly with bacteria, actinomycetes and yeast fungi, are not, however yet employed by lichenologists. The secondary metabolites in about 4500 lichens, around 33% of the species, have now been studied. Chemical data can therefore be used extensively in lichen systematics and also in discussions as to origins and relationships. The relatively sparse information on the components of most orders of non-lichenized fungi precludes meaningful comparisons with them (but see above), but within the lichen-forming fungi it has proved of value in delimiting higher categories. The Teloschistales are partly characterized by physcion and related anthraquinones (although some occur in representatives of other orders and groups, as noted above). The Peltigeraceae lack all depsidones and *meta*-depsides but do include terpene-containing species. The separation of the Ramalinaceae from the Usneaceae s.str. is supported by the presence of *meta*-depsides only in the former.

At the generic level, *Cetrelia* species lack caperatic acid while this occurs in all of *Platismatia*; here these and other supportive chemical data

Table 8.1 Comparison of selected chemical and anatomical characters separating *Alectoria* and some allied genera. Simplified from I.M. Brodo and D.L. Hawksworth (1977), *Opera bot.* **42**, 1–164.

	Alectoria	*Bryoria*	*Pseudephebe*	*Sulcaria*
Ascospores				
colour	brown	colourless	colourless	brown
septa	0	0	0	1 (− 3)
length (μm)	20–45	4–15	7–12	22–44
no. per ascus	2–4	8	8	(6 −) 8
Anatomy				
cortex	decomposing	usually smooth	smooth	uneven
pseudocyphellae	conspicuous	variable	absent	elongate
Chemistry				
usnic acids	usually present	absent	absent	absent
atranorin	absent	frequent	absent	frequent
orcinol depsides	frequent	rare	absent	absent
orcinol depsidones	frequent	absent	absent	rare
β-orcinol depsidones	absent	common	absent	present
pulvinic acid derivatives	absent	rare	absent	present

correlate with other characters (spore size and hypothecium height) to substantiate retention at the generic level. Usage in support of generic separation is now commonplace; a further example is provided in *Alectoria* and allied genera (table 8.1), where many correlations have been established.

The placement of a species in which no ascomata are known within a genus on the basis of its anatomy and chemistry is often possible (see section 1.5). The confidence in chemistry has been confirmed in many cases when sexual stages have subsequently been discovered. Species pairs (see section 3.8) characteristically have identical chemistries. However, it must be stressed that many of the lichen products are distributed through a wide range of genera, especially within the Lecanorales.

Controversy has surrounded the usage of chemical characters at the species level ever since Nylander used C (see above) to distinguish between two species of *Parmelia* in 1866. The discovery of different chemistries has often led to an appreciation of the importance of previously overlooked features, as in *Parmelia subrudecta* with lecanoric acid and a pale tan undersurface, and *P. borreri* with the corresponding tridepside gyrophoric acid and a black undersurface; these species also have different distributions. In some cases the differences are more subtle, as with the types of soralia and ecology in *Parmelia baltimorensis* and *P. caperata* in North America (see section 6.16). The *Cladonia chlorophaea* group presents particular problems as there appear to be correlations of the different chemical types with soredial size and colour and ecology in some parts of their range while these are less clear in others. The term *'chemotype'* has been introduced for chemical races within morphologically uniform lichens which have no, or undetermined, taxonomic significance.

In some cases only ecological or distributional tendencies appear to be correlated with chemical differences, and in this case the rank of variety may be used, as with *Thamnolia vermicularis* (var. *vermicularis*, thamnolic acid, predominates in the Southern Hemisphere; var. *subuliformis*, baeomycesic and squamatic acids, predominates in the Northern Hemisphere). In *Pseudevernia furfuracea*, a depsidone physodic acid chemotype (var. *furfuracea*) has a somewhat more southerly distribution than that with the corresponding depside olivetoric acid (var. *ceratea; P. olivetorina*) in Europe but in this case intermediate chemistries are known and some further compounds have been found to occur in low concentrations.

In the 1960s and early 1970s there was a tendency amongst lichen taxonomists, especially in North America, to recognize almost all chemical variants as separate species. This not only made identification difficult for those without chemical facilities but in some cases even obscured relationships that were present, as with *Ramalina siliquosa* where it is now recognized that there are two morphologically separate species (*R. cuspidata* and *R. siliquosa*) within each of which parallel chemotypes occur, not six species separated by their chemistries.

In a few instances, improved chemical techniques have revealed a type of chemical variation (*chemosyndromic*) in which one or two major constituents are accompanied by low amounts of several biogenetically related compounds; these major compounds can become the dominant ones in morphologically very similar lichens (table 8.2).

More importance is attached to replacements of one compound by another, especially if these are biogenetically distinct. The presence or absence of a single compound is generally regarded as of secondary importance, indeed they are often mentioned as *accessory substances*. An exception is provided by colour variants marked by the presence or absence of single pigmented compounds, where the chemotypes have been traditionally treated as species (e.g. *Parmeliopsis ambigua* and *P. hyperopta*).

However, in *Cladonia leucophaea* and *Haematomma ocholeucum* (see

Table 8.2 Major (**M**) and minor (*m*) medullary compounds identified in the 15 species of *Cetrelia*. (From C.F. Culberson and W.L. Culberson (1977), *Syst. Bot.* **1**, 325–339)

	Alectoronic	Anziaic	α-Collatolic	4-O-Demethylglomelliferic	4-O-Demethylimbricaric	4-O-Demethylmicrophyllinic	Divaricatic	Glomelliferic	Imbricaric	Loxodellic	4-O-Methylolivetoric	4-O-Methylphysodic	Microphyllinic	Olivetoric	Perlatolic	Physodic
C. alaskana		m				m	m	M	m						m	
C. braunsiana	m		m									M				m
C. cetrarioides		m						m	m		m			m	M	
C. chicitae	m		m									M				m
C. collata		m			m	m	m	M	m					M		
C. davidiana		m					m						M			
C. delavayana		m						m	m		m			M		
C. isidiata	M		m	m				m	m		m			m	m	
C. japonica							m					m	M			
C. monochorum		m			m	m	m	M	m						m	
C. nuda	m		m									M				m
C. olivetorum		m					m							M		
C. pseudolivetorum		m	m				m							M		
C. sanguinea	M		m	m				m	m		m				m	
C. sinensis						m	m	M	m						m	

section 6.16) there are correlations of the usnic acid deficient race with distributional and ecological differences respectively so the colour variants are retained as varieties. In other cases the pigment-deficient chemotypes are more sporadic, as in *Caloplaca verruculifera*, *Candelariella medians* and *Evernia prunastri*, or can occur as sectors in a single thallus as reported for *Caloplaca sinapisperma* and *Haematomma ventosum*.

8.5 Chemical hybrids

In the *Hypotrachyna* group of *Parmelia*, there are indications that the evolution of some extant groups of species may have involved the formation of hybrids between other types also still represented (figure 8.5). A further example is *Bryoria nadvornikiana* with the morphology of one section of the genus (*Divaricatae*) and chemistry of another (*Implexae*). If ancient hybridizations occurred, an entirely reasonable hypothesis for

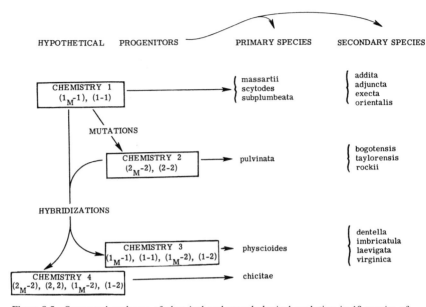

Figure 8.5 Suggested pathway of chemical and morphological evolution in 19 species of *Parmelia* subgenus *Hypotrachyna*. Adapted from C.F. Culberson and M.E. Hale (1973), *Brittonia* **25**, 162–173.

sexually reproducing organisms, there is every reason to suppose that hybridization occurs today. Possible examples are very rare specimens of *Alectoria imshaugii* with additional alectoronic acid from *A. sarmentosa*, one intermediate in morphology and with the chemistry of both *A. imshaugii* and *A. vancouverensis*, and ones with the morphology of *A. vancouverensis* and the chemistry of *A. sarmentosa*. This group of species is sympatric and the representatives of all form apothecia in western North America where these cases were discovered during examinations of substantial collections of these lichens.

Some apparent hybrids could, however, be 'mechanical' rather than 'genetical' in origin arising from propagules of different species growing from the same point and sharing a single photobiont to form a thallus with hyphae from both parents. Such a growing together of thalli has been documented in *Cladia* and could be more widespread than generally assumed. The correct interpretation of rare aberrant specimens that are chemical (and/or morphological) hybrids must remain uncertain in the absence of experimental data.

8.6   Role of lichen products

Flowering plants probably do not form complex secondary metabolites without a specific purpose, although these roles may not always be recognized. This may well be so in the case of lichens, but their function has not been critically examined. As *Trebouxia* isolates appear intolerant of excessive illumination, pigmented cortical compounds have been interpreted as having a role in protecting the photobiont from too much light. The production of usnic acid in the field and in isolated mycobionts has been shown to increase with illumination in a variety of macrolichens and this is also the case of physcion in *Xanthoria parietina* where the thallus colour varies from bright orange-red in exposed sunny sites to greyish yellow in shade (see table 5.7).

However, the same relationship has recently been demonstrated with colourless products in *Cladonia stellaris*, and not all common cortical compounds are pigmented (e.g. the colourless atranorin).

Lichens as a group are, however, remarkably resistant to attack by fungi, bacteria (see below) and arthropods. Indeed most of the fungi found on lichens are not common saprophytes but belong to genera and species restricted to the peculiar chemical environment of a lichen thallus (see section 1.4). The sparse reports of mould saprophytes in lichens all appear to arise from species which do not produce secondary metabolites, for example the records of *Aspergillus candidus* and *Trichothecium roseum* overgrowing endolithic pyrenocarpous lichens on limestone. There are limited studies which indicate that lichens can inhibit some fungi in pure culture (see section 6.8) and *Cladonia stellaris* can reduce the growth of *Pinus* seedlings by restricting the growth of mycorrhizal fungi, but more work in this area is needed to clarify the position. Grass seedlings may also be inhibited by *Peltigera* species.

Defence against browsing arthropods, gastropods and even mammals may be afforded by lichen products. This is especially so in the case of pulvinic acid derivatives, for example 150 mg of the vulpinic acid-containing *Letharia vulpina* would provide the LD_{50} level for a 50 g mouse. This lichen ('wolf's moss') has been used as a poison mixed with powdered glass for wolves in Lapland. Terpenes may also have a defensive function, and terpene-containing lichens tend to be less-grazed than many others in subtropical Stictaceae.

Chlorinated xanthones, especially frequent in coastal species, could also have a role in enabling lichens to survive in a chloride-rich site. An ecological role for secondary metabolites, other than oxalic acid, in the weathering of rock under saxicolous lichens does not appear to be of major importance (see section 6.4).

8.7   Antibiotic properties and medicinal effects

In the 1950s and 1960s, many macrolichens were screened for antibacterial and anti-actinomycete activity, and about half of those studied have been found to be effective against Gram-positive bacteria. Usnic acids are especially active and have tumor-inhibiting, antihistamine, spasmolytic and virucidal properties, as well as being active against Gram-positive bacteria and streptomycetes; they are used in seven commercially available antiseptic creams, such as 'Evosin' and 'Usno'. Erythrin from *Roccella montagnei* is still used in the preparation of erythritol tetranitrate, a drug sometimes used in the treatment of angina. Compounds with methylene-lactone groups may also have value as anti-cancer drugs (e.g. pro-tolichesterinic acid). The mycobiont wall carbohydrates lichenin and isolichenin also have anti-tumour activity, the latter has α-glucans and so does not cause the liver hardening that can arise from lichenin, which has β-glucan units. The medical possibilities of lichen compounds clearly merit further study.

In mediaeval times several lichens were used in herbal remedies, for example *Lobaria pulmonaria* ('tree lungwort') in the treatment of tuberculosis. *Cetraria islandica* ('Iceland moss') is still sold, usually in pastilles, for the relief of lung diseases and catarrh in Europe. A few lichens may have hallucinogenic effects, and have been used as mixers with tobacco.

Allergic dermatitis responses can occur in man, especially woodcutters or those working on lichen-covered fruit trees. Usnic and fumarprotocetraric acids appear to be particularly active in this regard, but only a minute proportion of people handling lichens become affected. Usnic acid is similar to psoralin, causing photosensitivity as it converts light to chemical energy with resultant rashes.

8.8   Dyeing

Lichens have been used as a source of dyestuffs at least since classical Greek times. Best-known are the red or purple colours obtained from species 'fermented' with ammonia. *Ochrolechia tartarea* and *O. androgyna* were exploited commercially in Glasgow in the early nineteenth century. These and other lichens, such as *Lasallia pustulata* and *Roccella montagnei*, can be broken up and 'dampened' with ammonia solution (1/3 full strength) and sealed in a container in which 2/3 of the volume is air. After one week the colour will develop and is a 'direct' purple dye for wool and silk, i.e. protein fibres. The colour is sensitive to pH and in a slightly different form has been used commercially to colour litmus paper. The substances in lichens that produce these colours are *para*-depsides, notably lecanoric acid (*Ochrolechia*), gyrophoric acid (*Lasallia pustulata*), and erythrin (*Roccella*). They are first hydrolysed by the ammonia and then undergo a complex series of chemical changes involving the co-operation of amino groups to form the orcin chromophore.

Lichens have also been used as a brown dye. Species of *Parmelia*, especially *P. omphalodes*, were collected from the Scottish Highlands and Islands to dye protein fibres (wool and perhaps silk) brown. This species contains the colourless salazinic acid, having an aldehyde group in the molecule. In some manner this acid reacts with the protein of the wool to form a pigment of indeterminate composition. Other lichen products containing aldehyde groups are also capable of dyeing wool brown to yellowish colours.

A knowledge of the chemistry of lichens makes it possible to select species likely to dye either of these two basic colours. However, other sources of these colours are available without endangering slow-growing lichen species.

8.9 Perfumes

Evernia prunastri ('oak moss') and *Pseudevernia furfuracea* ('tree moss') are used extensively in the manufacture of perfumes. Perfumes require an ingredient to promote persistence on the skin which does not itself evaporate rapidly. Lichen extracts are used for this purpose in some perfumes that are compatible with the sweet 'mossy' smell lichen extracts impart. The extract addition amounts to 1–12% of the finished perfume. There is no precise knowledge of the identity of the substances in the lichen that produce the desired effect generally available outside the industry, but they are soluble in organic solvents and oily in nature. The scented component is a very small proportion of the total (0.04%); the remainder including camphor, borneol, cineole, naphthalene, geraniol, thusone, citronellol and breakdown products of lichen metabolites.

CHAPTER NINE

ENVIRONMENTAL MONITORING

The intimate physiological relationship between lichen thalli and the environment, the perennial nature of lichens and their sensitivity to disturbance, dependence on nutrients and chemicals not derived from their substratum, an ability to concentrate compounds from weak solutions, and also the range of species with different requirements and sensitivities, means that lichens act as continuous monitors of the environment. An appreciation of their qualities as biological monitors, and the study of the parameters limiting the occurrence of particular species, has led to their use as indicators of a variety of environmental factors. The literature on this subject is now vast, comprising in excess of 1000 scientific papers and books, and only selected aspects can be reviewed here. Several more comprehensive reviews are listed under the list of further reading for this chapter and those wishing to undertake project work in this field are advised to consult them.

9.1 Air pollution

Some naturalists in the early nineteenth century had already recognized that there was a link between the sparseness of the lichen vegetation in towns and air pollution. In 1866, as a result of studies in Paris, Nylander suggested that lichens could act as indicators of the intensity of that pollution. Dramatic declines in the lichen flora in the early decades of the present century stimulated further work and by the 1930s three and four zones of lichen impoverishment were being mapped around pollution centres. Scant physical measurements of ambient pollution levels prevented correlations with lichen distributions being made until the late 1960s and it is only since these were established that it has been possible to utilize lichens more fully in biological monitoring of air pollution.

The most important gaseous air pollutant affecting lichens in cooltemperate regions is sulphur dioxide. Acid rain accentuates the direct effects of this gas (see below) by reducing the pH of bark to the extent that whole communities can be replaced by ones characteristic of more acid bark (see section 7.7). The extent of the changes can be dramatic and there have been substantial modifications in the lichen flora over almost the whole European plain in the course of the last 60–100 years (see section 7.7).

The difference in sensitivity of lichens to air pollution arises from the interplay of two series of factors, avoidance and tolerance. Avoidance includes the reduction of assimilation through limiting wettability (as through the powdery-sorediate surface and large amounts of fumaroprotocetraric acid in *Lecanora conizaeoides*), thallus structure (particularly the formation of narrow convex lobes), and reductions of toxic ions through the pH and buffering capacity of both the thallus and the substratum. Tolerance when toxic compounds reach the living protoplasts is often of secondary importance. When avoidance procedures are overcome, for example by artificially saturating a thallus in the laboratory, some of the apparently more 'pollution-tolerant' lichens appear to be much more sensitive than others which are observed to be more susceptible in field situations.

Mean sulphur dioxide levels as low as 30 μg m^{-3} can have marked effects on some lichens in the field. Such low concentrations are difficult to handle experimentally both in the field and in the laboratory; most laboratory studies have therefore been short-term and with higher sulphur dioxide concentrations. Photosynthesis has been found to be the most sensitive physiological process, although respiration is also affected.

The effects of sulphur dioxide on photosynthesis are manifold. At low concentrations, carbon dioxide uptake may be irreversibly inhibited, apparently by the competitive inhibition of ribulose diphosphate carboxylase. At higher concentrations irreversible damage occurs exemplified by the leakage of potassium ions through damaged membranes, and chlorophyll damage resulting in changes in spectral characteristics (and even breakdown to phaeophytin). Sudden exposure to high concentrations can even result in the algal cells becoming bleached. Lichen metabolites can be affected by compounds leaking from damaged cells to produce coloured pigments (e.g. red from salazinic acid in *Parmelia sulcata*). In general, experimental treatments have corresponded to sensitivities in the field, but it must be remembered that what takes place in the field is a complex association of physiological effects, thallus morphology and ecological distribution on a microclimatic scale. There is also a complex relationship between time and concentration which is not entirely understood, and the physiological state (especially water content and hence metabolic activity) when exposed can also be crucial. The absence of a protective cuticle may be a key factor in explaining the sensitivity of lichens (and mosses) to gaseous pollutants.

The activity of sulphur dioxide in causing damage is very dependent on pH. At lower pH values (below 4), it is very toxic, but at pH values above 5 it is much less so, leading to lichen colonization on alkaline substrata (e.g. limestone and asbestos-cement) in quite polluted areas (see table 9.2). The dependence of pH is probably due to the ionizing characteristics of sulphur dioxide in solution. Below pH 4 free sulphur dioxide exists in aqueous

Table 9.1 Zone scale for the estimation of mean winter sulphur dioxide levels in England and Wales using corticolous lichens. Adapted from D.L. Hawksworth and F. Rose (1970), *Nature* **227**, 145–148.

Zone	Moderately acid bark	Mean Winter $SO_2(\mu g\,m^{-3})$
0	Epiphytes absent	?
1	*Desmococcus viridis* s.l. present but confined to the base	170
2	*Desmococcus viridis* s.l. extends up the trunk; *Lecanora conizaeoides* present but confined to the bases	about 150
3	*Lecanora conizaeoides* extends up the trunk; *Lepraria incana* becomes frequent on the bases	about 125
4	*Hypogymnia physodes* and/or *Parmelia saxatilis*, or *P. sulcata* appear on the bases but do not extend up the trunks. *Hypocenomyce scalaris*, *Lecanora expallens* and *Chaenotheca ferruginea* often present	about 70
5	*Hypogymnia physodes* or *P. saxatilis* extends up the trunk to 2.5 m or more; *P. glabratula*, *P. subrudecta*, *Parmeliopsis ambigua* and *Lecanora chlarotera* appear; *Calicium viride*, *Chrysothrix candelaris* and *Pertusaria amara* may occur; *Ramalina farinacea* and *Evernia prunastri* if present largely confined to the bases; *Platismatia glauca* may be present on horizontal branches	about 60
6	*P. caperata* present at least on the base; rich in species of *Pertusaria* (e.g. *P. albescens*, *P. hymenea*) and *Parmelia* (e.g. *P. revoluta* (except in NE), *P. tiliacea*, *P. exasperatula* (in N); *Graphis elegans* appearing; *Pseudevernia furfuracea* and *Bryoria fuscescens* present in upland areas	about 50
7	*Parmelia caperata*, *P. revoluta* (except in NE), *P. tiliacea*, *P. exasperatula* (in N) extend up the trunk; *Usnea subfloridana*, *Pertusaria hemisphaerica*, *Rinodina roboris* (in S) and *Arthonia impolita* (in E) appear	about 40
8	*Usnea ceratina*, *Parmelia perlata* or *P. reticulata* (S and W) appear; *Rinodina roboris* extends up the trunk (in S); *Normandina pulchella* and *U. rubicunda* (in S) usually present	about 35
9	*Lobaria pulmonaria*, *L. amplissima*, *Pachyphiale cornea*, *Dimerella lutea*, or *Usnea florida* present; if these absent crustose flora well developed with often more than 25 species on larger well lit trees	under 30
10	*L. amplissima*, *L. scrobiculata*, *Sticta limbata*, *Pannaria* spp., *Usnea articulata*, *U. filipendula* or *Teloschistes flavicans* present to locally abundant	'Pure'

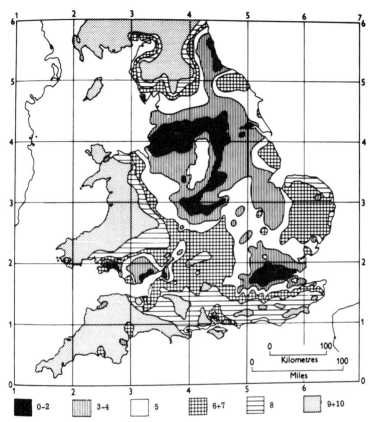

Figure 9.1 Approximate limits of lichen zones (table 9.1) in England and Wales. From D.L. Hawksworth and F. Rose (1976), *Lichens as Pollution Monitors*, Arnold, London.

solution while above that pH it is mainly as the bisulphite ion (HSO_3^-; bactericidal) or in alkaline solutions as sulphite ions (SO_3^{2-}).

Acid rain, made acid by whatever source (sulphur dioxide, nitrogen oxides, etc.) is damaging also in its own right by lowering bark pH (see section 7.7). The pH of oak bark may be changed from 5.5–6.5 to 3.0–4.5 in this way, and resultant changes to a more oligotrophic lichen flora can be seen in the Black Forest and Ardennes particularly, where coniferous trees are also affected.

Lichen distributions around pollution sources can be investigated by a variety of methods. The individual mapping of species is often very informative but also extremely time-consuming, and for the survey of large areas zone scales which use a variety of species (actually more often communities adapted to substrata of particular pH values) have proved more successful. The scheme developed by Hawksworth and Rose in 1970 has been widely used in the British Isles and other parts of western Europe; the extensive information on sulphur dioxide levels in the British Isles enabled their scale to be calibrated (table 9.1) and it has been used to produce national (figure 9.1) and more local maps. The method is of especial value when combined with information from recording gauges which not only confirm calibration but are often located in areas where the pollution has eliminated almost all the lichens. Quantitative methods have also been devised but are both time-consuming and include elements of subjectivity, as with the Index of Atmospheric Purity (IAP). In addition the accumulation of sulphur in the thalli of species can be compared and mapped in a similar way to that adopted in studies of heavy metal fallout (see section 9.2). Lichens on trees have proved of most value in sulphur dioxide monitoring, but a great deal of complementary and substantive information can be gained from lichens on other substrata. In all these approaches, particular attention has to be paid to standardization in view of the ameliorating effects of substrata and other factors; a single species can also change its substratum with increasing pollution levels (table 9.2).

Where lichens or trees are absent, transplants of lichen thalli have also been used successfully in the monitoring of both sulphur dioxide and fluoride pollution, especially in Germany and Norway. Bark discs or twigs bearing lichens can be fastened to boards fastened on poles.

Table 9.2 Biological scale for the status of *Lecanora muralis* compared with distance from the centre of Leeds, pH of rainwater and sulphur dioxide levels from M.R.D. Seaward (1977), in *Lichenology: Progress and Problems* (D.H. Brown *et al.*, eds.), Academic Press, London and New York, 323–357.

Distance from Leeds city centre in 1970 (miles)	Status of Lecanora muralis	Mean annual SO_2 $(\mu g/m^{-3})$	Rainwater pH
0–1.5	Absent	> 240	4.4–4.7
1.5–2.5	On asbestos-cement tile roofs	200–240	4.7–4.9
2.5–3.5	On asbestos-cement tile and sheet roofs	170–200	4.9–5.1
3.5–5.5	On asbestos-cement roofs, cement, concrete and mortar	125–170	5.1–5.5
Over 5.5	On asbestos-cement roofs, cement, concrete and siliceous wall capstones	< 125	Over 5.5

As with any biological system, care is needed in interpreting results. Rather little is known of the time taken to establish lichen communities, and growth rates are also small (see Chapter 4). Where pollution levels are increasing lichen response is usually rapid, but when these are ameliorating—a situation almost unknown when the main correlations were established in the late 1960s—a time-lag was to be expected and does indeed occur. The extent of any lag will vary according to the proximity of propagules of species, and the time taken for adjustments in the pH of the substrata. There factors mean that under ameliorating conditions of sulphur dioxide pollution considerable caution is needed in the interpretation of field observations. Nevertheless, cases of reinvasion are now being documented and our understanding of the factors involved is starting to increase. For example, in the London area and the West Yorkshire conurbation, where sulphur dioxide levels have declined substantially over the last decade, ages for the establishment of colonies have been calculated from their growth rates (figure 9.2).

Patterns may also be disrupted by particularly high fumigations for short periods, for example if the plume from a tall chimney comes to earth

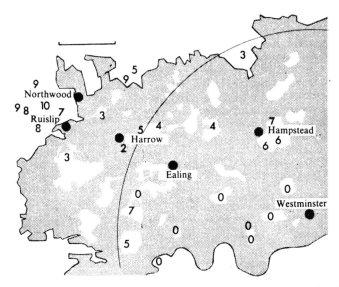

Figure 9.2   Lichen re-colonization in north-west London. Numbers indicate the minimum age (y) of foliose lichens on trees calculated from thallus sizes and growth rates. The arc is a 16-km radius based on Trafalgar Square. Scale = 5 km. From C.I. Rose and D.L. Hawksworth (1981), *Nature* **289**, 289–292.

unexpectedly when conditions are moist (under dry conditions when the lichens are inactive effects would normally be insignificant).

Sulphur dioxide is not the only pollutant to affect lichens. Fluorides are especially toxic but much less widespread in their effects, occurring mainly in association with brickworks or aluminium smelters. Zones comparable to those developing around sulphur dioxide sources occur but the order of the species affected may differ. *Buellia pulverea* was described from the vicinity of an aluminium smelter in Scotland but is known from sites also affected by other pollutants elsewhere in the British Isles.

Oxidant air pollutants, including ozone and peroxyacetylnitrate (PAN), are the major components of photochemical smog which is of major importance in hot and dry urban and industrialized areas such as the Los Angeles region. Studies within the last three years have demonstrated that lichens may be adversely affected in the field; earlier fumigations in the laboratory, utilizing ozone, PAN and nitrogen oxides separately had been inconclusive, even showing increased photosynthesis in one investigation, and the situation still needs critical study. Synergistic effects may be of paramount importance and could also involve the sulphur dioxide invariably present in urban areas. A recent study in Norway has also found lichens of value in monitoring polychlorinated biphenyls (PCB's) near a busy road intersection by examination of the PCB content of thalli of *Ramalina duriaei*.

Soot particles themselves have little effect on lichens. Comparative studies of the lichen vegetation where both sulphur dioxide and particulates have been measured provide correlations only with levels of the former (table 9.3).

Table 9.3. Comparison of mean smoke and sulphur dioxide levels and lichen zones at selected sites in England. Compiled from data in B.W. Ferry, M.S. Baddeley and D.L. Hawksworth (eds.) (1973), *Air Pollution and Lichens* (Athlone Press of the University of London).

Site	Winter mean 1967–70		Lichen zone 1967–70 (Table 9.1)
	smoke ($\mu g\,m^{-3}$)	sulphur dioxide ($\mu g\,m^{-3}$)	
Leicester 14	89	175	0–1
Buxton 2	17	126	3
Plymouth 13	97	82	3–4
Abbots Ripton 1	30	61	5
Torquay 3	33	32	8
Weymouth	13	22	9

9.2 Heavy metals

Lichens are able to concentrate heavy metals from dilute solutions (see section 5.8), depositing them on the surfaces of hyphae or in cell walls outside the protoplast. Assays of the metal contents of lichens have been used extensively in the monitoring of heavy metal fallout from smelters in Europe and North America. *Peltigera* species have proved especially useful for such studies and can persist at massive metal levels, withstanding for example even 90 000 ppm of iron. Zones based on different ranges of concentrations can be drawn to indicate the fallout patterns; these are normally related to the direction of the prevailing winds. Metal contents tend to fall off rapidly with the distance from the pollution source (figure 9.3).

9.3 Radionuclides

Radionuclides derived from nuclear detonations in the atmosphere and on land in the 1950s and early 1960s resulted in the fallout of radioactive isotopes such as ^{137}Cs and ^{90}Sr. These became concentrated in lichens eaten by reindeer and caribou (see section 6.12) which in turn were eaten by humans, giving cause for concern. Regular programmes to monitor radionuclide levels in lichens were established involving the annual examination of *Cladonia* subgenus *Cladina* species, for example *C. rangiferina*, where each internode represents one season of growth. With the cessation of nuclear testing on the surface, levels have declined

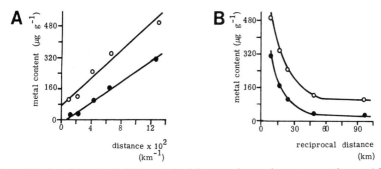

Figure 9.3 Iron (o) and nickel (•) contents of *Stereocaulon* sp. along a transect from a nickel smelter in Sudbury, Ontario. *A*, As a function of the reciprocal distance. *B*, As a function of distance. Iron contents $\times 10^{-1}$. Based on data of F.D. Tomassini. Adapted from D.H.S. Richardson and E. Nieboer (1981), *Endeavour* **5**, 127–133.

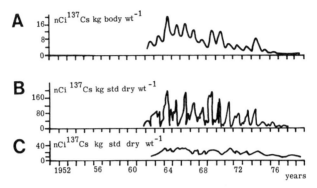

Figure 9.4   Caesium-137 concentrations ($nCi\ kg^{-1}$) at Anaktuvuk Pass, Alaska, 1962–79. *A*, Eskimos. *B*, Caribou flesh. *C*, Lichens. Adapted from W.C. Hanson (1982), *Health Phys.* **42**, 433–447.

significantly over the last decade and are now well below ones hazardous to human health (figure 9.4)—in addition Eskimos now eat fewer caribou.

9.4   Ecological continuity

Comparative studies of the lichens present in numerous woodlands and ancient parks in the British Isles established that some lichens were only found in sites which had had mature trees present for many centuries. Indeed, those richest in such species have histories of maintenance as forest or deer park extending back to some of the earliest written records of the eleventh century and have been derived from portions of fragments of the primeval forests established after the decline of the last ice sheets.

These *old forest indicator lichens* require ecological continuity of mature trees to persist and have only limited powers of dispersal so that they only invade new sites with extreme difficulty. The retention of a closed tree canopy and a humid woodland interior environment appear to be less important than the trees themselves as most of these can persist as relicts on ancient trees in exposed parkland sites. The Revised Index of Ecological Continuity (RIEC) gives a numerical indication of the extent of ecological continuity in a woodland:

$$RIEC = \frac{n}{20} \times 100$$

The value n is the number of species from a list of old forest indicator

Table 9.4 Species used to calculate the Revised Index of Ecological Continuity (RIEC). Adapted from F. Rose (1976), in *Lichenology: Progress and Problems* (D.H. Brown *et al.*, eds.), Academic Press, London and New York, 279–307.

Arthonia vinosa	*L. laetevirens*	*P. horizontalis*
Arthopyrenia cinereopruinosa	*L. pulmonaria*	*Pertusaria pupillaris*
Catillaria atropurpurea	*L. scrobiculata*	*Porina leptalea*
C. sphaeroides	*Nephroma laevigatum*	*Pyrenula chlorospila*
Dimerella lutea	*Pachyphiale cornea*	*Rinodina isidioides*
Enterographa crassa	*Pannaria conoplea*	*Stenocybe septata**
Haematomma elatinum	*Parmelia crinita*	*Sticta limbata*
Lecanactis lyncea	*P. reddenda*	*S. sylvatica*
L. premnea	*Parmeliella corallinoides*	*Thelopsis rubella*
Lobaria amplissima	*Peltigera collina*	*Thelotrema lepadinum*

* Not lichenized.

lichens (table 9.4) which are present in the site. This presentation is used in preference to simply *n* to provide sufficient representation of regional differences in the British Isles.

Many groups of organisms include representatives which are confined to woodlands which have an ecological continuity of mature trees, for example flowering plants, macrofungi, beetles and spiders. However, lichens, because they are perennial and non-mobile, afford a rapid method of assessing the relative importance of woodland sites. The technique, developed by Rose in 1974–76, has found wide applications in site assessment in the British Isles but is limited to areas where the levels of sulphur dioxide are low. The method will fail to identify ancient coppiced woodlands (unless there has been a continuity in the standard trees) and has so far been devised only for deciduous woodlands.

There is little doubt, however, that the approach also holds true for other woodland types and over many geographical regions. For example in the pine woodlands of central Scotland *Alectoria sarmentosa, Bryoria capillaris, B. lanestris, Cavernularia hultenii, Hypocenomyce friesii* and *Usnea extensa* are amongst species which appear to behave as old forest indicators. A detailed investigation into the sites in which *Usnea longissima* grows in Sweden published in 1981 established that it was restricted to spruce forests with long histories of ecological continuity and in addition that its powers of dispersal (mainly by lengths of thallus torn off by the wind and branch movements) were so limited that it was often restricted to individual trees; fragments were mainly within 2 m and exceptionally 3 m of a tree with the species. In North America other examples may be recognized, for example *Lobaria* and *Ocellularia* species in ancient sugar maple stands (but not planted woodlands) in Quebec and Ontario.

It is reasonable to presume that restricted powers of dispersal limit the distribution of some saxicolous and terricolous lichens in a parallel manner but this hypothesis is yet to be tested.

9.5 Lichenometry

By measuring the radial growth rate of circular lichen thalli it is possible to calculate the probable age of the thallus. This can be used to estimate the date a recently exposed substratum was first colonized. The dating of substratum exposure in this way is called *lichenometry* and has been used widely to date the exposure of moraines of glaciers over several hundred years. The most widely used lichen species is *Rhizocarpon geographicum*, as it is usually abundant in such habitats, although other equally slow-growing species can also be employed. The rate of growth of *Rhizocarpon geographicum* varies over its lifespan, with very small thalli growing more slowly, medium-sized thalli growing at a constant and more rapid rate ('grand phase'), with old large thalli growing more slowly again. Estimating the age of thalli of different sizes has been achieved by (a) measuring the size of thalli on substrata of known age, and (b) measuring the growth rate of thalli of different sizes. For quick reference in the field, it has been found convenient to have a rough guide as to the age of thalli of different sizes. This is refered to as the '*lichen factor*', and refers to the size of a 100-year-old thallus of *R. geographicum*. The 'lichen factor' varies considerably with geographical location, and estimates range from 2–40 mm in Greenland to 20–87 mm for Scotland.

In using lichen growth to determine the age of substrata, various practical points must be considered.

(a) Species must be determined accurately. Some groups or genera (e.g. *Rhizocarpon*) are taxonomically complex and species identification requires care.

(b) The substratum should be uniform. The rate of growth of a lichen species may be affected by variations in rock type or chemistry.

(c) The habitat needs to be uniform. In lichenometric moraine dating, the microclimate of the rock surface may change with distance from the retreating glacier. The slope of the rock surface and nutrient input from bird droppings may also have an effect.

(d) The rate of growth of thalli of different sizes should be checked for the site under investigation. This can be achieved by comparing photographs taken one year or more apart.

Lichenometry has been used to date other substrata, such as stone monuments, walls, and rockfalls on cliffs. Although it has proved a useful technique, there can be considerable errors which need to be carefully assessed in any particular application. The technique has great potential but still requires placing on a firmer theoretical base.

9.6 Potential uses

The potential of lichens as monitors of environmental factors remains to be exploited in additional fields, but insufficient information is currently available to make this practicable now. Topics where there are indications that they may be of value include pollution of freshwater streams and lakes, agricultural chemicals, indicating chemical composition of rocks, and metal prospecting.

FURTHER READING

General

Ahmadjian, V. and Hale, M.E., eds. (1974) ['1973'] *The Lichens*. Academic Press, New York etc.

Brown, D.H., Hawksworth, D.L. and Bailey, R.H., eds. (1976) *Lichenology: Progress and Problems*. Academic Press, London etc.

Hale, M.E. (1983) *The Biology of Lichens*. 3rd edn., Edward Arnold, London.

Hawksworth, D.L. (1974) *Mycologist's Handbook. An introduction to the principles of taxonomy and nomenclature in the fungi and lichens*. Commonwealth Mycological Institute, Kew.

Hawksworth, D.L., Sutton, B.C. and Ainsworth, G.C. (1983) *Ainsworth & Bisby's Dictionary of the Fungi (Incli̇ding the Lichens)*. 7th edn., Commonwealth Mycological Institute, Kew. [Diagnoses of orders, families etc.; generic names; glossary of lichenological terms.]

Henssen, A. and Jahns, H.M. (1973) ['1974'] *Lichenes. Eine Einführüng in die Flechtenkunde*. G. Thieme, Stuttgart.

Richardson, D.H.S. (1975) *The Vanishing Lichens. Their history, biology and importance*. David & Charles, Newton Abbot etc.

Seaward, M.R.D., ed. (1977) *Lichen Ecology*. Academic Press, London etc.

Identification

Brodo, I.M. (1981) Lichens of the Ottawa region. *Syllogeus* **29**, 1–37. [Also in French.]

Cannon, P.M., Hawksworth, D.L. and Sherwood-Pike, M.A. (1984) *The British Ascomycotina: an annotated checklist*. Commonwealth Mycological Institute, Kew.

Dobson, F. (1981) *Lichens. An illustrated guide*. 2nd edn., Richmond Publishing, Richmond-upon-Thames.

Duncan, U.K. and James, P.W. (1970) *An Introduction to British Lichens*. T. Buncle, Arbroath. [Richmond Publishing, Richmond-upon-Thames.]

Hale, M.E. (1979) *How to Know the Lichens*. 2nd edn., Wm. C. Brown, Dubuque, Iowa.

Hale, M.E. and Culberson, W.L. (1970) A fourth checklist of the lichens of the continental United States and Canada. *Bryologist* **73**, 499–543.

Hawksworth, D.L. (1977) 'A bibliographic guide to the lichen floras of the world' in *Lichen Ecology* [see general references], 437–502. [Includes floras by country and state, also checklists and regional monographs.]

Hawksworth, D.L., James, P.W. and Coppins, B.J. (1980) Checklist of British lichen-forming, lichenicolous and allied fungi. *Lichenologist* **12**, 1–115. [Also available separately.]

Jahns, H.M. (1980) *Farne. Moose. Flechten. Mittel-, Nord- und Westeuropas*. BLV Verlagsgesellschaft, München etc. [English version (1983) *Ferns, Mosses & Lichens of Britain and Northern and Central Europe*. Collins, London.]

Krog, H., Østhagen, H. and Tønsberg, T. (1980) *Lav Flora. Norske busk- og bladlav*. Universitetsforlaget, Oslo etc. [Also with supplementary keys in English.]

Moberg, R. and Holmåsen, I. (1982) *Lavar. En fälthandbok*. Interpublishing, Stockholm.

Ozenda, P. and Clauzade, G. (1970) *Les Lichens. Étude biologique et flore illustrée*. Masson, Paris.

142

Poelt, J. (1969) *Bestimmungschlüssel europäischer Flechten.* J. Cramer, Lehre. [and Vězda, A. (1977, 1981) *Erganzungsheft I, II, Bibliotheca lich.* 9, 16.]
Wirth, V. (1980) *Flechtenflora.* Eugen Ulmer, Stuttgart.

Chapter 1

Ahmadjian, V. (1967) A guide to the algae occurring as lichen symbionts: isolation, culture, cultural physiology, and identification. *Phycologia* **6**, 127–160.
Ahmadjian, V. (1970) The lichen symbiosis: its origin and evolution. *Evolut. Biol.* **4**, 163–184.
Ahmadjian, V. (1980) 'Separation and artificial synthesis in lichens', in *Cellular Interactions in Symbiosis and Parasitism*, eds. C.B. Cook, P.W. Pappas and E.D. Rudolph, Ohio State University Press, Columbus, 3–29.
Ahmadjian, V. (1982) Algal/fungal symbioses. *Progr. Phycol. Res.* **1**, 179–233.
Ainsworth, G.C. (1976) *Introduction to the History of Mycology.* Cambridge University Press, Cambridge.
Eriksson, O. (1983) Outline of the ascomycetes—1983. *Systema Ascomycetum* **2**, 1–37.
Hawksworth, D.L. (1973) Some advances in the study of lichens since the time of E.M. Holmes. *Bot. J. Linn. Soc.* **67**, 3–31.
Hawksworth, D.L. (1978) 'The taxonomy of lichen-forming fungi: reflections on some fundamental problems' in *Essays in Plant Taxonomy*, ed. H.E. Street, Academic Press, London etc., 211–243.
Hawksworth, D.L. (1982) Secondary fungi in lichen symbioses: parasites, saprophytes and parasymbionts. *J. Hattori bot. Lab.* **52**, 357–366.
Hawksworth, D.L. and Seaward, M.R.D. (1977) *Lichenology in the British Isles 1565–1975.* Richmond Publishing, Richmond-upon-Thames.
Hildreth, K.C. and Ahmadjian, V. (1981) A study of *Trebouxia* and *Pseudotrebouxia* isolates from different lichens. *Lichenologist* **13**, 65–86.
Kohlmeyer, J. and Kohlmeyer, E. (1979) *Marine Mycology—The Higher Fungi.* Academic Press, New York etc.
Law, R. and Lewis, D.H. (1983) Biotic environments and the maintenance of sex—some evidence from mutualistic symbioses. *Biol. J. Linn. Soc.* **20**, 249–276.
Richardson, D.H.S. (1975) *The Vanishing Lichens. Their history, biology and importance.* David & Charles, Newton Abbot etc.
Smith, A.L. (1921) *Lichens.* Cambridge University Press, Cambridge. [Republished 1975, Richmond Publishing, Richmond-upon-Thames.]
Vobis, G. and Hawksworth, D.L. (1981) 'Conidial lichen-forming fungi' in *The Biology of Conidial Fungi*, eds. G.T. Cole and B. Kendrick, Academic Press, New York etc., **1**, 245–273.

Chapter 2

Ahmadjian, V. (1982) Algal/fungal symbioses. *Progr. Phycol. Res.* **1**, 179–233.
Beltman, H.A. (1978) Vegetative Strukturen der Parmeliaceae und ihre Entwicklung. *Bibliotheca lich.* **11**, 1–193.
Brodo, I.M. and Richardson, D.H.S. (1978) Chimeroid associations in the genus *Peltigera. Lichenologist* **10**, 157–170.
Hale, M.E. (1973) Fine structure of the cortex in the lichen family Parmeliaceae viewed with the scanning-electron microscope. *Smithson. Contr. bot.* **10**, 1–92.
Hale, M.E. (1976) 'Lichen structure viewed with the scanning electron microscope', in *Lichenology: Progress and Problems*, [see general references] 1–15.

Hale, M.E. (1981) Pseudocyphellae and pored epicortex in the Parmeliaceae: their delimitation and evolutionary significance. *Lichenologist* **13**, 1–10.

Hannemann, B. (1973) Anhangsorgane der Flechten. *Bibliotheca lich.* **1**, 1–123.

Hawksworth, D.L. (1973) 'Ecological factors and species delimitation in the lichens' in *Taxonomy and Ecology*, ed. V.H. Heywood, Academic Press, London etc., 31–69.

Jahns, H.M. (1974) 'Anatomy, morphology and development' in *The Lichens*, [see general references] 3–58.

James, P.W. and Henssen, A. (1976) 'The morphological and taxonomic significance of cephalodia' in *Lichenology: Progress and Problems*, [see general references] 27–77.

Ozenda, P. (1963) Lichens. *Handb. Pflanzenanatomie* **6**(9), 1–199.

Poelt, P. (1974) 'Systematic evaluation of morphological characters' in *The Lichens*, [see general references] 91–115.

Weber, W.A. (1977) 'Environmental modification and lichen taxonomy' in *Lichen Ecology*, [see general references] 9–29.

Chapter 3

Bellemère, A. and Letrouit-Galinou, M.A. (1981) 'The lecanoralean ascus: an ultrastructural preliminary study' in *Ascomycete Systematics*, ed. D.R. Reynolds, Springer, New York, 54–67.

Bowler, P.A. and Rundel, P.W. (1975) Reproductive strategies in lichens. *Bot. J. Linn. Soc.* **70**, 325–340.

Eriksson, O. (1981) The families of bitunicate ascomycetes. *Opera bot.* **60**, 1–240.

Henssen, A. (1981) 'The lecanoralean centrum' in *Ascomycete Systematics*, ed. D.R. Reynolds, Springer, New York, etc., 138–234.

Henssen, A. and Jahns, H.M. (1973) ['1974'] *Lichenes. Eine Einführung in die Flechtenkunde.* G. Thieme, Stuttgart.

Honegger, R. (1978–83) The ascus apex in lichenized fungi I–IV. *Lichenologist* **10**, 47–67; **12**, 157–172; **14**, 205–217; **15**, 57–71.

Honegger, R. (1984) Scanning electron microscopy of the contact site of conidia and trichogynes in *Cladonia furcata*. *Lichenologist* **16**, 11–19.

Poelt, J. (1972) Die taxonomische Behandlung von Artenpaaren bei den Flechten. *Bot. Notiser* **125**, 77–81.

Poelt, J. (1974) 'Systematic evaluation of morphological characters' in *The Lichens*, [see general references] 91–115.

Vobis, G. (1980) Bau und Entwicklung der Flechten-Pycnidien und ihrer Conidien. *Bibltheca lich.* **14**, 1–141.

Vobis, G. and Hawksworth, D.L. (1981) 'Conidial lichen-forming fungi' in *The Biology of Conidial Fungi*, ed. G.T. Cole and B. Kendrick, Academic Press, New York etc., **1**, 245–273.

Chapter 4

Armstrong, R.A. (1976) The influence of the frequency of wetting and drying on the radial growth of three saxicolous lichens in the field. *New Phytol.* **77**, 719–724.

Bailey, R.H. (1976) 'Ecological aspects of dispersal and establishment in lichens' in *Lichenology: Progress and Problems*, [see general references] 215–247.

Bubrick, P., Galun, M. and Frensdorff. A. (1981) Proteins from the lichen *Xanthoria parietina* which bind to phycobiont cell walls. Localisation in the intact lichen and cultured mycobiont. *Protoplasma* **105**, 207–211.

Fisher, P.J. and Proctor, M.C.F. (1978) Observations on a season's growth in *Parmelia caperata* and *P. sulcata* in South Devon. *Lichenologist* **10**, 81–90.

Hale, M.E. (1974) 'Growth' in *The Lichens*, [see general references] 473–492.

Hill, D.J. (1981) The growth of lichens with special reference to the modelling of circular thalli. *Lichenologist* **13**, 265–287.

Pyatt, F.B. (1974) 'Lichen propagules' in *The Lichens*, [see general references] 117–145.

Rhoades, F.M. (1977) Growth rates of the lichen *Lobaria oregana* as determined from sequential photographs. *Can. J. Bot.* **55**, 2226–2233.

Seaward, M.R.D. (1976) 'The performance of *Lecanora muralis* (Schreb.) Rabenh. in an urban environment' in *Lichenology: Progress and Problems*, [see general references] 323–357.

Topham, P.B. (1977) 'Colonization, growth, succession and competition' in *Lichen Ecology*, [see general references] 31–68.

Chapter 5

Ahmadjian, V. (1967) *The Lichen Symbiosis*. Blaisdell, Waltham, Mass.

Ahmadjian, V. (1982) Algal/fungal symbioses. *Progr. phycol. Res.* **1**, 179–233.

Bewley, J.D. (1979) Physiological aspects of desiccation tolerance. *A. Revs. Plant Physiol.* **30**, 195–238.

Bewley, J.D. and Krochko, J.E. (1982) 'Desiccation-tolerance' in *Encyclopedia of Plant Physiology* N.S. 12B (*Physiological Plant Ecology II, Water Relations and Carbon Assimilation*, eds O.L. Lange, P.S. Nobel, C.B. Osmond, and H. Ziegler), Springer, Berlin-New York, 325–378.

Brown, D.H. and Beckett, R. (1984) Uptake and effect of cations on lichen metabolism. *Lichenologist* **16**, 173–188.

Cowan, D.A. Green, T.G.A. and Wilson, A.T. (1979) Lichen metabolism I. The use of tritium labelled water in studies of anhydrobiotic metabolism in *Ramalina celastri* and *Peltigera polydactyla*. *New Phytol.* **82**, 489–503.

Crittenden, P.D. (1983) 'The role of lichens in the nitrogen economy of subarctic woodlands: nitrogen loss from the nitrogen fixing lichen *Stereocaulon paschale* during rainfall' in *Nitrogen as an Ecological Factor*, eds J.A. Lee, S. McNeill, and I.H. Rorison, Blackwell, Oxford, 43–68.

Farrar, J.F. (1976) 'The lichen as an ecosystem: observation and experiment' in *Lichenology: Progress and Problems*, [see general references] 385–406.

Farrar, J.F. (1984) 'Symbiosis between algae and fungi' in *Handbook of Nutrition and Food*, CRC Press, Cleveland (in press).

Hill, D.J. (1976) 'The physiology of the lichen symbiosis' in *Lichenology: Progress and Problems*, [see general references] 457–496.

Honegger, R. (1984) Cytological aspects of the mycobiont –phycobiont relationship in lichens. *Lichenologist* **16**, 111–127.

Kershaw, K.A. (1983) The thermal operating-environment of a lichen. *Lichenologist* **15**, 191–207.

Kershaw, K.A. (1984) Seasonal photosynthetic capacity changes: a provisional mechanistic interpretation. *Lichenologist* **16**, 145–171.

Kershaw, K.A. (1984) *Physiological Ecology of Lichens*. Cambridge University Press, Cambridge (in press).

Lange, O.L., Geiger, I.L. and Schulze, E.D. (1977) Ecological investigations on the lichens of the Negev Desert. V A model to simulate net photosynthesis and respiration of *Ramalina maciformis*. *Oecologia* (Berl.) **28**, 247–259.

Lange, O.L. and Matthes, U. (1981) Moisture-dependent CO_2 exchange of lichens. *Photosynthetica* **15**, 555–574.

MacFarlane, J., Kershaw, K.A. and Webber, M.R. (1983) Physiological environmental interactions in lichens. XIII Phenotypic differences in the seasonal pattern of net photosynthesis in *Cladonia rangiferina*. *New Phytol.* **94**, 217–233.

Matthes, U. and Feige, B. (1983) 'Ecophysiology of lichen symbioses' in *Encyclopedia of Plant Physiology* N.S. 12C. *Physiological Plant Ecology III. Responses to the Chemical and Biological Environment*, eds O.L. Lange, P.S. Nobel, C.B. Osmond, and H. Ziegler, Springer, Berlin and New York, 423–467.

Millbank, J.W. (1982) The assessment of nitrogen fixation and throughput by lichens. III Losses of nitrogenous compounds by *Peltigera membranacea, P. polydactyla* and *Lobaria pulmonaria* in simulated rainfall episodes. *New Phytol.* **92**, 229–234.

Nieboer, E.N., Richardson, D.H.S. and Tomassini, F.D. (1978) Mineral uptake and release by lichens. An overview. *Bryologist* **81**, 226–246.

Smith, D.C. (1975) Symbiosis and the biology of lichenised fungi. *Symp. Soc. exp. Biol.* **29**, 373–405.

Smith, D.C. (1981) 'Mechanisms of nutrient movement between the lichen symbionts' in *Cellular Interactions in Symbiosis and Parasitism*, eds. C.B. Cook, P.W. Pappas, and E.D. Rudolph, Ohio State University Press, Columbus 197–227.

Stewart, W.D.P., Rowell, R., and Rai, A.N. (1980) 'Symbiotic nitrogen fixing cyanobacteria' in *Nitrogen Fixation*, eds W.D.P. Stewart & J.R. Gallon, Academic Press, London & New York 239–277.

Chapter 6

Barkman, J.J. (1958) *Phytosociology and Ecology of Crypogamic Epiphytes*. Van Gorcum, Assen.

Brodo, I.M. (1974) 'Substrate ecology' in *The Lichens*, [see general references] 401–441.

Fletcher, A. (1980) 'Marine and maritime lichens of rocky shores: their ecology, physiology and biological interactions' in *The Shore Environment*, eds. J.H. Price, D.E.G. Irvine and W.F. Farnham, Academic Press, London etc., **2**, 789–842.

James, P.W., Hawksworth, D.L. and Rose, F. (1977) 'Lichen communities in the British Isles: a preliminary conspectus' in *Lichen Ecology*, [see general references] 295–413.

Kershaw, K.A. (1984) *Physiological Ecology of Lichens*. Cambridge University Press, Cambridge (in press).

Richardson, D.H.S. (1975) *The Vanishing Lichens*. David & Charles, Newton Abbot etc.

Seaward, M.R.D. ed. (1977) *Lichen Ecology*. Academic Press, London etc.

Syers, J.K., and Iskander, I.K. (1974) 'Pedogenetic significance of lichens' in *The Lichens*, [see general references] 225–248.

Wilson, M.J. and Jones, D. (1983) 'Lichen weathering of minerals: implications for pedogenesis' in *Residual Deposits: surface related weathering processes and materials* (ed. R.C.L. Wilson), Blackwell Scientific Publications, Oxford, 5–12.

Chapter 7

Coppins, B.J. (1976) 'Distribution patterns shown by epiphytic lichens in the British Isles' in *Lichenology: Progress and Problems*, [see general references] 249–278.

Culberson, W.L. (1972) Disjunctive distributions in the lichen-forming fungi. *Ann. Mo. bot. Gdn.* **59**, 165–173.

Galloway, D.J. (1979) 'Biogeographical elements in the New Zealand lichen flora' in *Plants and Islands*, ed. B. Bramwell, Academic Press, London etc., 201–224.

Hawksworth, D.L. (1977) 'A bibliographic guide to the lichen floras of the world' in *Lichen Ecology*, [see general references] 437–502.

Hawksworth, D.L., Coppins, B.J., and Rose, F. (1974) 'Changes in the British lichen flora' in

The Changing Flora and Fauna of Britain, ed. D.L. Hawksworth, Academic Press, London etc., 47–78.

Jørgensen, P.M. (1983) Distribution patterns of lichens in the Pacific region. *Aust. J. Bot., Suppl. Ser.* **10**, 43–66.

Seaward, M.R.D. and Hitch, C.J.B. (1982) *Atlas of the Lichens of the British Isles*, vol. 1. Institute of Terrestrial Ecology, Cambridge.

Sipman, H. (1983) A monograph of the lichen family Megalosporaceae. *Bibliotheca lich., Vaduz* **18**, 1–241.

Tehler, A. (1983) The genera *Dirina* and *Roccellina* (Roccellaceae). *Opera bot.* **70**, 1–86.

Thomson, J.W. (1972) Distribution patterns of American arctic lichens. *Can. J. Bot.* **50**, 1135–1156.

Chapter 8

Brodo, I.M. (1978) Changing concepts regarding chemical diversity in lichens. *Lichenologist* **10**, 1–11.

Culberson, C.F. (1969) *Chemical and Botanical Guide to Lichen Products*. University of North Carolina Press, Chapel Hill.

Culberson, C.F. (1970) Supplement to 'Chemical and Botanical Guide to Lichen Products.' *Bryologist* **73**, 177–377.

Culberson, C.F. (1972) Improved conditions and new data for the identification of lichen products by a standardized thin-layer chromatographic method. *J. Chromat.* **72**, 113–125.

Culberson, C.F., Culberson, W.L. and Johnson, A. (1977) *Second Supplement to 'Chemical and Botanical Lichen Products'*. American Bryological and Lichenological Society, St. Louis.

Culberson, C.F. and Ahmadjian, V. (1980) Artificial reestablishment of lichens, II. Secondary products of resynthesized *Cladonia cristatella* and *Lecanora chrysoleuca*. *Mycologia* **72**, 90–109.

Culberson, C.F., Culberson, W.L., and Johnson, A. (1983) Genetic and environmental effects on growth and production of secondary compounds in *Cladonia cristatella*. *Biochem. Syst. Ecol.* **11**, 77–84.

Culberson, W.L. and Culberson, C.F. (1970) A phylogenetic view of chemical evolution in the lichens. *Bryologist* **73**, 1–31.

Hale, M.E. (1983) *The Biology of Lichens*. 3rd edn., Edward Arnold, London.

Hawksworth, D.L. (1976) 'Lichen chemotaxonomy' in *Lichenology: Progress and Problems*, [see general references] 139–184.

Huneck, S. (1971) Chemie und Biosynthese der Flechtenstoffe. *Fortschr. Chem. org. Naturstoffe* **29**, 209–306.

Kok, A. (1966) A short history of the orchil dyes. *Lichenologist* **3**, 248–272.

Mosbach, K. (1974) 'Biosynthesis of lichen substances' in *The Lichens*, [see general references] 523–546.

Richardson, D.H.S. (1975) *The Vanishing Lichens. Their history, biology and importance*. David & Charles, Newton Abbot etc.

Rundel, P.W. (1978) The ecological role of secondary lichen substances. *Biochem. Syst. Ecol.* **6**, 157–170.

Sheard, J.W. (1978) The taxonomy of the *Ramalina siliquosa* species aggregate (lichenized ascomycetes). *Can. J. Bot.* **56**, 915–938.

Vartia, K.O. (1974) 'Antibiotics in lichens' in *The Lichens*, [see general references] 547–561.

Walker, F.J. and James, P.W. (1980) A revised guide to microchemical techniques for the identification of lichen products. *Bull. Br. Lichen Soc.* **46** (*Suppl.*) 13–29.

Chapter 9

Ferry, B.W., Baddeley, M.S. and Hawksworth, D.L., eds. (1973) *Air Pollution and Lichens.* Athlone Press of the University of London, London.

Hanson, W.C. (1982) ^{137}Cs co ncentrations in northern Alaskan eskimos 1962–79: effects of ecological, cultural and political factors. *Health Phys.* **42**, 433–447.

Hawksworth, D.L. (1971) Lichens as litmus for air pollution: an historical review. *Internat. J. environ. Stud.* **1**, 281–296.

Hawksworth, D.L. and Rose, F. (1976) *Lichens as Pollution Monitors.* [Studies in Biology no. 66.] Edward Arnold, London.

Lock, W.W., Andrews, J.T. and Webber, P.J. (1979) *A Manual for Lichenometry.* British Geomorphological Research Group, Geoabstracts, Norwich.

Nash, T.H. III (1976) Lichens as indicators of air pollution. *Naturwissenschaften* **63**, 364–367.

Nieboer, E. and Richardson, D.H.S. (1981) 'Lichens as monitors of atmospheric deposition' in *Atmospheric Pollutants in Natural Waters*, ed. S.J. Eisenreich, Ann Arbor. Science, Ann Arbor, 339–388.

Proctor, M.C.F. (1983) Sizes and growth rates of thalli of the lichen *Rhizocarpon geographicum* on the moraines of the glacier de Valsorey, Valais, Switzerland. *Lichenologist* **15**, 249–261.

Richardson, D.H.S. and Nieboer, E. (1981) Lichens and pollution monitoring. *Endeavour* **5**, 127–133.

Rose, F. (1976) 'Lichenological indicators of age and environmental continuity in woodlands' in *Lichenology: Progress and Problems*, [see general references] 279–307.

Seaward, M.R.D. (1980) 'The use of lichens as bioindicators of ameliorating environments' in *Bioindikation auf der Ebene der Individuen*, eds. R. Schubert and J. Schuh. Martin Luther Universität, Halle-Wittenberg, 17–23.

Seaward, M.R.D. (1982) 'Lichen ecology of changing urban environments' in *Urban Ecology*, ed. R. Bornkamm, J.A. Lee and M.R.D. Seaward, Blackwell Scientific Publications, Oxford etc., 181–189.

Türk, R., Wirth, V. and Lange, O.L. (1974) CO_2-Gaswechsel-Untersuchungen zur SO_2-Resistenz von Flechten. *Oecologia* **15**, 33–64.

Webber, P.J. and Andrews, J.T. (1973) Lichenometry: a commentary. *Arctic Alpine Res.* **5**, 295–302.

Wetmore, C.M., ed. (1984) *Biomonitoring Air Quality with Lichens and Bryophytes.* Benjamin/Cummings Publishing, Minnesota (in preparation).

Postscript

Ahmadjian, V., and Jacobs, J.B. (1983) 'Algal–Fungal relationships in lichens: recognition, synthesis and development', in *Algal Symbiosis*, ed. L.J. Goff, Cambridge University Press, 147–172.

Santesson, R. (1984) *The Lichens of Norway and Sweden.* Swedish Museum of Natural History, Uppsala and Stockholm.

Index

This Index should be used in conjunction with the table of Contents (pp. vii-ix). Pages on which definitions of technical terms appear are indicated in *italic* type; for the most part terms are listed in the singular form.